Taunton's **COMPLETE ILLUSTRATED** *Guide to*

Jigs & Fixtures

SANDOR NAGYSZALANCZY

The Taunton Press

The Taunton Press
Inspiration for hands-on living®

The Taunton Press, Inc., 63 South Main Street, PO Box 5506, Newtown, CT 06470-5506
e-mail: tp@taunton.com

EDITOR: Paul Anthony
DESIGN: Lori Wendin
LAYOUT: Susan Lampe-Wilson, Lori Wendin
ILLUSTRATOR: Melanie Powell
PHOTOGRAPHERS: Sandor Nagyszalanczy

LIBRARY OF CONGRESS CATALOGING-IN-PUBLICATION DATA:

Nagyszalanczy, Sandor.
  Taunton's complete illustrated guide to jigs & fixtures / Sandor Nagyszalanczy.
      p. cm.
  Includes indexes.
  ISBN-13: 978-1-56158-770-4
  ISBN-10: 1-56158-770-2
  1. Woodworking tools. 2. Jigs and fixtures. I. Title: Complete illustrated guide to jigs & fixtures. II. Title.
  TT186.N3365 2006
  684'.08--dc22
                                2005021522

Printed in Italy
10 9 8 7 6 5 4 3 2 1

To BOU DOU, my ever faithful companion

Acknowledgments

WHEN I WROTE MY FIRST BOOK ON WOODWORKING almost a dozen years ago (*Woodshop Jigs & Fixtures*), I was fortunate to have an entire cadre of insightful, experienced, and creative people supporting me. Fortunately, many of those same people have given me their guidance and made significant contributions to the development of this project. First and foremost, I must thank my publishing rock of Gibraltar, Helen Albert, without whom I'd be lost in a sea of scribbled pages. I can't express adequate gratitude to her for sticking by me all these years and always generously providing her expertise in both woodworking and publishing (along with a healthy dose of compassion and understanding). Also kudos to Jennifer Peters, Helen's right-hand gal, who has done at least a thousand favors for me and never dropped the ball once. Thanks too, to Julie Hamilton, for keeping a steady hand on the reins of this project. High praise goes to Paul Anthony, my editor on this project, for his keen eye for detail (and nose for bull puckey). This book is far better than it would have been without his consummate skills and patience.

With regard to the materials and supplies needed for creating jigs for this book, I'd like to thank several companies and the individuals who contributed. A strong dose of my appreciation goes to Leonard Lee and Wally Wilson at Lee Valley Tools, who graciously gave me tools, hardware, and jig-building materials used in many of the photo-essays. I have the greatest respect for Lee Valley's commitment not only to making high-quality woodworking tools and supplies but to improving woodworking and making it a safer and more joyful pursuit. I share a similar admiration for WoodHaven, and thank Brad Witt for his many years of support and generosity. Thanks too to: Rob Johnstone, editor of *Woodworker's Journal* magazine; Bill Perrizo at Rockler, Paul Fitzmaurice at Imre Communications (Ridgid & Ryobi power tools), Christian Oltzscher of Festool USA, and Lisa Agostoni at Freud America, for all their support and kind assistance.

Among my local cadre of woodworking colleagues, warm thanks to furniture maker extraordinaire Roger Heitzman, who not only supplied some of the jig-building materials used in this book but generously shared his considerable woodworking expertise and creative insight. Thanks too to Michael Dresdner for finishing supplies and information, and to Jerry Bowden for supplying me with a handsome turning and so many great flea-market tool finds. Gratitude to my friend and sometimes-assistant Steven Robins. Also, a big fat "yee-haa" to the terrific DJs at KPIG radio,107.5 in Freedom, Calif., for keeping me company during many a lonely night as I torched the midnight oil in my photo studio.

Finally, no project could be completed without the patience and understanding of the loved ones closest to you; they share your joy on good days, empathize with your pain on bad days, and make countless sacrifices and compromises in the name of creation. Thanks to my dear parents, Lorant and Maria, for their undying love and support. And a whole-hearted thanks to my wife, Ann MacGregor Gibb, who often rescued me from being buried under a pile of jigs with the buoyancy of her smile.

Contents

PART TWO Woodworking Jigs · 44

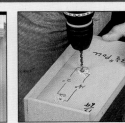

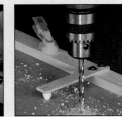

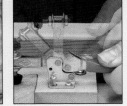

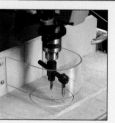

About Your Safety

WORKING WITH WOOD IS INHERENTLY DANGEROUS. Using hand or power tools improperly or ignoring safety practices can lead to permanent injury or even death. Don't try to perform operations you learn about here (or elsewhere) unless you're certain they are safe for you. If something about an operation doesn't feel right, don't do it. Look for another way. We want you to enjoy the craft, so please keep safety foremost in your mind whenever you're in the shop.

Please note that in some of the photos in this book, guards have been removed for clarity. For your own safety, always use the appropriate guards when operating tools and machines.

How to Use This Book

FIRST OF ALL, this book is meant to be used, not put on a shelf to gather dust. It's meant to be pulled out and opened on your bench when you need to do a new or unfamiliar technique. So the first way to use this book is to make sure it's near where you do woodworking.

In this book, we've included a wide variety of techniques to fit these needs.

To find your way around the book, you first need to ask yourself two questions: What result am I trying to achieve? What tools do I want to use to accomplish it?

To organize the material, we've broken the subject down to two levels. "Parts" are major divisions of this class of techniques. "Sections" contain related techniques.

Within sections, techniques and procedures that create a similar result are grouped together, usually organized from the most common way to do it to methods requiring specialized tools or a larger degree of skill. In some cases, the progression starts with the method requiring the most basic technology and then moves on to alternative methods using other common shop tools and finally to specialized tools.

The first thing you'll see in a part is a group of photos keyed to a page number. Think of this as an illustrated table of contents. Here you'll see a photo representing each section in that part, along with the page on which each section starts.

Each section begins with a similar "visual map," with photos that represent major groupings of techniques or individual techniques. Under each grouping is a list of the step-by-step essays that explain how to do the methods, including the pages on which they can be found.

Sections begin with an "overview," or brief introduction, to the methods described therein. Here's where you'll find important general information on this group of techniques, including any safety issues. You'll also read about specific tools needed for the operations that follow and how to build jigs or fixtures needed for them.

The step-by-step essays are the heart of this book. Here a group of photos represents the key steps in the process. The accompanying text describes the process and guides you through it, referring you back to the photos. Depending on how you learn best, either read the text first or look at the photos and drawings; but remember, they are meant to work together. In cases where there is an alternative step, it's called out in the text and the visual material as a "variation."

The "VISUAL MAP" tells you where to locate the essay that details the operation you wish to do.

A "SECTION" groups related processes together.

The "OVERVIEW" gives you important general information about the group of techniques, tells you how to build jigs and fixtures, and provides advice on tooling and safety.

For efficiency, we've cross-referenced redundant processes or steps described in another related process. You'll see yellow "cross-references" called out frequently in the overviews and step-by-step essays.

When you see this symbol ⚠, make sure you read what follows. The importance of these safety warnings cannot be overemphasized. Always work safely and use safety devices, including eye and hearing protection. If you feel uncomfortable with a technique, don't do it, try another way.

At the back of the book is an index to help you find what you're looking for in a pinch. There's also list of further reading to help you brush up on how to use tools and keep them sharp, as well as some general references on design.

Finally, remember to use this book whenever you need to refresh your memory or to learn something new. It's been designed to be an essential reference to help you become a better woodworker. The only way it can do this is if you make it as familiar a workshop tool as your favorite bench chisels.

–The editors

"STEP-BY-STEP ESSAYS" contain photos, drawings, and instructions on how to do the technique.

"WARNINGS" tell you specific safety concerns for this process and how to address them.

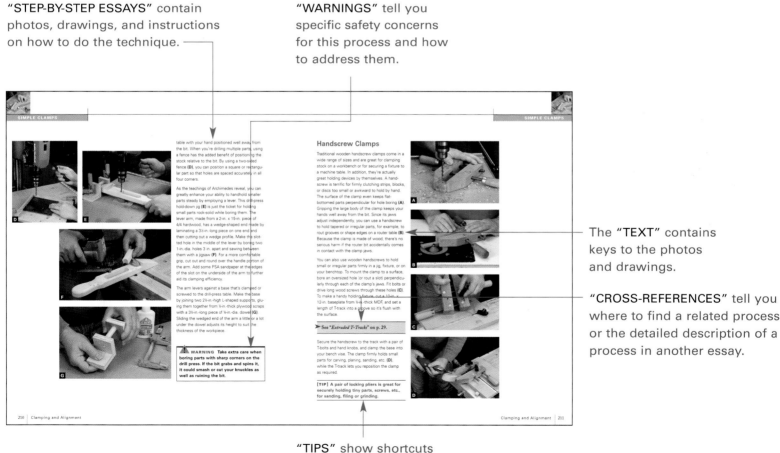

The "TEXT" contains keys to the photos and drawings.

"CROSS-REFERENCES" tell you where to find a related process or the detailed description of a process in another essay.

"TIPS" show shortcuts and smart ways to work.

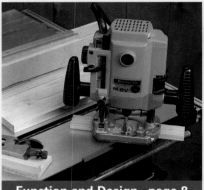

PART ONE

Design and Materials

BEFORE YOU CAN BEGIN building your own jigs and fixtures, you need to know what to build and how to build it. A good way to start is to explore basics of jig design, appropriate materials and hardware, and good construction practices. Whether you plan to build jigs exactly as shown in this book or design your own customized devices, you'll benefit from learning the functional approach to what jigs do, and how to go about choosing or creating jigs that suit your tools and particular work style. Once you've nailed down a design, you must choose from a wide assortment of wood and synthetic materials and hardware (fasteners, fittings, extrusions, etc.) to make a jig that's accurate and functional. Finally, you must employ good construction practices to make your jig sturdy and durable. You'll accomplish that by using strong joinery and solid woodworking practices, along with a few tricks gleaned from the machinist's trade.

Function and Design

FINDING AN EFFECTIVE JIG OR FIX-TURE for a woodworking operation can be as elusive and time-consuming as designing a great piece of furniture, a sailboat, or a ukulele. It takes solid woodworking knowledge and some problem-solving skills, with a good dose of inspiration thrown in. Even a moderately complex setup can pose numerous choices: What tool or machine is best for the operation? Should the jig move the wood over the machine or guide the machine past the wood? How does the workpiece need to be referenced and clamped? Should the jig be adjustable? Most of the woodworkers I've met delight in solving challenging jigging problems. But tackling these problems requires a basic understanding of jig function and design.

Generally speaking, jigs are the devices that help us cut, shape, drill, and sand parts quickly and accurately, while fixtures are devices that hold parts firmly during machining and assembly (see the photo on the facing page). Used separately or together, these essential woodworking aids help us do everyday tasks like cutting boards to accurate length and width, as well as complex operations like pattern routing parts or sawing three-dimensional shapes. While we all own at least a few commercially made jigs (miter gauges, router edge guides, etc.), most woodworkers create their own jigs. In this section, we'll examine the function of jigs and fix-

tures and how they help us perform woodworking tasks more easily, accurately, and safely. We'll also explore some factors to consider when choosing the best jig for a particular job, as well as how to customize jigs to better suit your needs.

Why Use Jigs?

With all the tools in an average workshop, why do we need jigs and fixtures at all? For starters, some tools are nearly useless without them: Imagine ripping or crosscutting on a tablesaw without a rip fence and a miter gauge (see the top photo on p. 10). By using a variety of other jigs, you can greatly expand a tablesaw's basic repertoire and cut a variety of joinery, shape moldings and panels, and more. Whether basic or complex, jigs and fixtures are an essential part of many tool and machine operations. Here are some of the ways woodworkers make good use of jigs and fixtures:

Performing operations that are difficult or impossible to do freehand. Imagine cutting perfectly mitered ends on a picture frame without a jig to guide the cut, or assembling a loose stack of bevel-edged staves into a perfectly symmetrical curved cabinet door without the proper form. Want impossible? How about sawing out a perfect circle without a circle guide? (See the bottom photo on p. 10.)

Machining parts with precision and repeatable accuracy. Examples are cutting a

dozen rails for a face frame exactly 13⅝ in. long, or drilling holes at a precise 33-degree angle. Jigs also help us do repetitive processes—say, cutting tenons on 100 frame members—with greater consistency and safety, which is precisely why jigs are an important part of production woodworking.

Expanding the capabilities of tools and machines. Fences let you cut raised panel edges on a tablesaw, or saw compound-curved parts on a bandsaw. The right jig can even take ordinary everyday hand tools and transform them for other uses. For example, removed from a regular bench plane, a plane blade serves as a useful cutter in a trimming jig used for mitering the ends of wood banding strips for decorative inlays (see the top photo on p. 11).

The Functional Approach

To help you fulfill your jigging and fixturing needs, Parts II and III of this book are filled with photo-essays that show you how to build complete jigs and fixtures to tackle literally hundreds of different woodworking jobs, from initial layout to final assembly. Instead of being arranged by the machine they're used with (jigs for the tablesaw, etc.), they're grouped by the kind of function they perform: marking and layout, fences and guides, sliding and pivoting jigs, tables and tops, portable power tool and hand tool jigs, and templates and patterns. In Part III I'll describe fixtures for part-holding, assembly, and lamination, as well as an assortment of jigging devices, including stops, positioning blocks, and clamps used to adapt and customize jigs.

Here's a compound-curve sawing jig on the bandsaw (rear), a pivoting arc-sawing jig for the bandsaw (left), and an edge-gluing fixture for making panels (right).

A good rip fence and miter gauge are essential jigs for ripping and crosscutting and hence are standard equipment with every tablesaw.

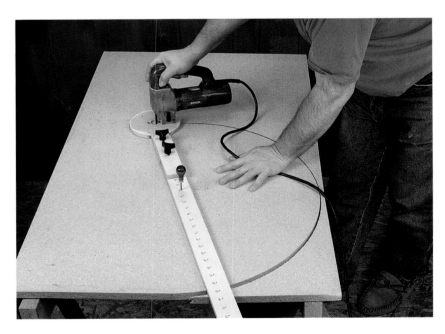

It takes more than a steady hand to guide a jigsaw when you're cutting out a perfect circular disc, and an adjustable circle jig is just what's needed.

The reason for the function-oriented organization is simple: There are usually at least several different ways to perform any particular operation, such as cutting a tenon, shaping a molding, or clamping up a panel. Before choosing one type of jig over another, it's important to consider a number of factors, including the machine to be used, the workpiece, and the ultimate versatility of the jig.

Which Machine to Use?

You can plow a groove with a dado blade on the tablesaw, with a straight bit in the router, or by handplaning with a rabbet plane (see the bottom photo on the facing page). The decision depends not only on which tools you do (or don't) own but on how easy a tool is to set up for the operation, or how accurately and cleanly it can do the job. Also, one jig setup may be a lot simpler and easier to build than another for a certain machine.

The Size of the Workpiece

Typically, it's easier to leave really big, heavy workpieces stationary and guide a portable tool to do the cutting, rather than try to move the large workpiece over a stationary machine. For example, you can cut up a large sheet of plywood or particleboard with a portable circular saw and fence guide (see the top photo on p. 12) instead of trying to wrestle the sheet over the top of your tablesaw.

Dedicated or Adjustable?

An adjustable jig can handle a wider range of operations than a jig that's dedicated to a single function. For example, an adjustable

An ordinary bench plane blade acts as the cutting tool in a guillotine-like trimming jig that cuts perfectly square ends or miters on decorative wood banding strips.

Grooves and rabbets can be cut using a router with an edge guide, a tablesaw with a rip fence, or a handplane with a depth stop.

tilting table on your drill press lets you bore holes at a variety of angles. However, adjustable jigs require more time, effort, and materials to build, must be adjusted before use, and can accidentally slip out of adjustment.

On the other hand, dedicated jigs are simpler to build and require little or no setup before use. For example, the adjustable taper jig, shown at the rear of the bottom photo on p. 12, handles a wide range of tapering jobs, but the dedicated taper jig shown at front uses a simple notched piece of plywood to taper all four sides of a chair leg. The dedicated jig is not only easier to make but can be reused in the future without the user having to remember particular adjustments. On the downside, you have to make a new dedicated jig for each and every job.

Safety

When choosing between jigs, always prefer the one that's safest. In most cases, that means choosing the machine and jig setup that lets you perform the operation without

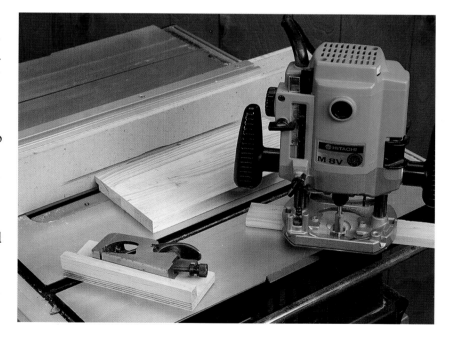

putting yourself in harm's way. For example, a mitering jig used with a radial-arm saw is great for cutting large frame members, but a small-parts cutting carriage used on the tablesaw is better for crosscutting small wood strips because it protects your fingers from

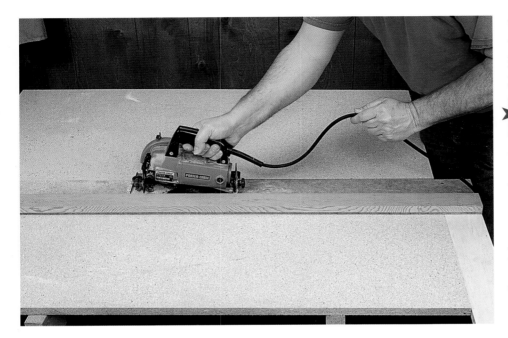

It's best to cut up a large plank or panel by leaving it stationary while using a portable circular saw guided by a fence jig.

Although both of these jigs are made for tapering parts on the tablesaw, the one at the rear adjusts for different tapers, while the jig in front is dedicated to sawing a particular taper.

the blade and prevents parts from being thrown. (You can make any jig safer by incorporating protective guards and dust collection, as shown in Sections 13 and 14.)

▶ See *"Small-Part Routing"* on p. 114.

Customizing Jigs

Although many jigs found in this book are ready to build and use as shown, there's usually a lot to gain by altering and adapting a jig's design to suit your machines as well as your particular needs. Changing the size of a jig or fixture, as well as customizing its construction and features, not only helps enhance its performance but enables the device to better suit your style of work. Here are some important things to consider:

Changing Dimensions

Very few jigs in this book are accompanied by dimensioned drawings. That's done on purpose, because the overall size of the jig you build is affected by the size of your machines as well as your workpieces. You can adjust the dimension of practically any jig part to suit your particular needs (see the top photo on the facing page). Once you've outlined the overall proportions, make a quick sketch and fill in the dimensions of all the necessary parts and hardware before you start to build.

Combining Features

Probably the most significant way to customize a jig to suit a particular application is to "mix and match" various components of different jigs to get the features and functionality you desire. Most jigs are already combinations of several functional elements.

For example, consider the part-shaping carriage shown in the bottom photo. The base of the jig is a template that guides the cut, with stops that position the workpiece, clamps that hold it securely during shaping, and handles that allow the user to work safely.

You can combine all manner of jig features to suit your needs. You can add stops to a fence on a cutoff device, or add a tilt table to a hold-down fixture to allow angled drilling. Likewise, a micro-adjuster added to a tapering jig will allow you to tweak angle settings. Sections 10 and 11 cover positioning, adjustment, and clamping devices—many of which aren't complete jigs or fixtures themselves, but are elements specifically designed for customizing jigs for particular applications. For example, mounting quick-action pneumatic clamps to a parts-trimming jig would allow a production woodworker to cut hundreds of parts a day much more quickly and efficiently, as shown in the top photo on p. 14.

Another customizing method is to build upon the jigs you already own. Instead of making a tablesaw box joint jig from scratch, you could add the necessary components to a regular miter gauge.

▶ See *"Box Joints"* on p. 101 and 178.

The universal sliding jig is designed to have different jig components bolt onto it so the jig can be used to bevel or raise panels, cut tenons and splines, taper legs, and more.

▶ See *"Universal Fence Jig"* on p. 104.

Most jigs, like these sliding tablesaw carriages, can be built bigger or smaller to suit the size of your machines and scale of your work.

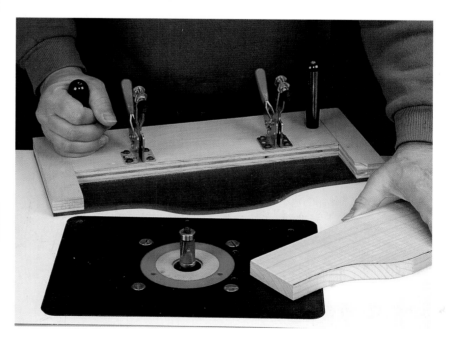

Adding a curved template to the base of this shaper carriage allows you to customize the jig to shape the edge of a workpiece as desired.

Changing Materials or Hardware

To make a jig easier to build, I have often described in the photo-essay the particular materials and hardware used. However, you can usually substitute other materials to suit the stock you have on hand, or to fulfill a particular purpose. For example, even though I recommend using quality sheet materials (MDF, plywood, etc.) for strong, long-lasting jigs, there's no reason you can't use particleboard for a jig or assembly fixture you plan to use only once.

Similarly, feel free to change jig hardware to suit your application. If you build a jig larger than shown or want to beef it up for heavy-duty use, use larger fasteners and hardware and stronger construction techniques as necessary. For instance, if you're building a really large tablesaw sled, using heavy-duty metal miter gauge slot bars will help the jig slide straighter and truer.

Multi-Machine Jigs

An excellent way to expand the functionality of a jig is to adapt its design and/or construction so it will work on more than one machine. A good example is the circle jig, shown and described at left.

The basic jig is made for cutting out round wheels, rings, and tabletops on a bandsaw. But with a couple of extra components, the same jig also works on the stationary disc sander for sanding the edges of rough-cut wood discs smooth and to exact diameter.

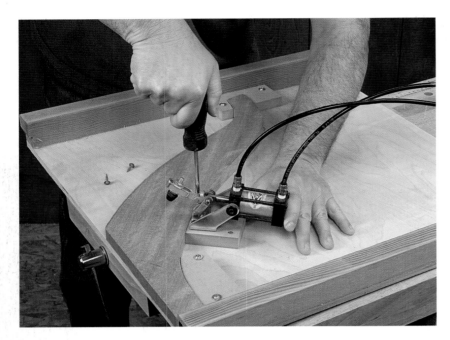

Mounting an air-powered clamp to this sliding parts-trimming carriage allows rapid clamping of parts for high-volume production work.

This circle jig was originally made for the bandsaw, but a few alterations adapt it for edge-trimming on a disc sander, making the jig more versatile.

Materials and Hardware

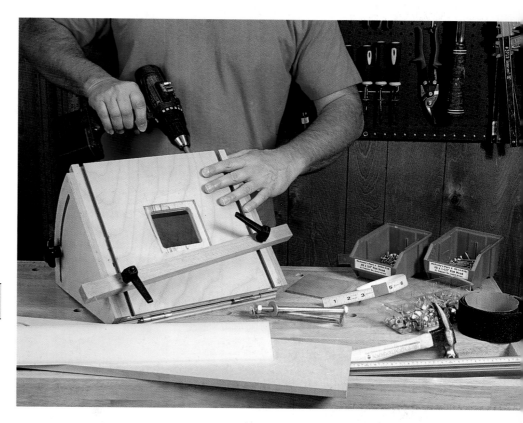

Y OU MIGHT BE CONTENT TO COBBLE UP a jig or fixture out of any old bits and pieces you find in your shop's scrap box and junk drawer. However, jigs made for everyday use definitely last longer and retain their accuracy better when built from high-quality materials using solid woodworking construction practices. Jigs built from quality materials and hardware are a pleasure to use and yield more accurate and consistent results than jury-rigged, makeshift setups, so they may actually save you time and trouble in the long run.

> ▶ See *"Tools and Techniques"* on pp. 33-34.

There is usually no single "right" material for any given jig you're building. But all materials—including solid wood, sheet goods, and plastics—have a range of applications for which they are best suited. Understanding the properties, strengths, and weaknesses of these jig-building materials will help you select the best choice for a particular application.

The same is true for the hardware items you'll need to use when assembling and outfitting your custom jigs and setups. Finding just the right piece of hardware when building a custom jig or fixture can save a lot of fuss and bother.

In addition to standard hardware items—screws, hinges, bolts, etc.—there's a ton of hardware specifically designed for wood-

working jigs and fixtures, including knobs and handscrews, miter guide bars, extruded tracks, self-stick measuring tapes, and more. Most of the hardware items described in this section are available at good hardware stores and home centers, or through mail order supply catalogs.

Selecting Materials

Whether you're making production fixtures for a giant cabinet factory or a small setup for a one-man shop, most materials have their jig-building niche. In the following

Using the right materials and hardware is the key to building strong, accurate jigs that hold up to the rigors of woodworking.

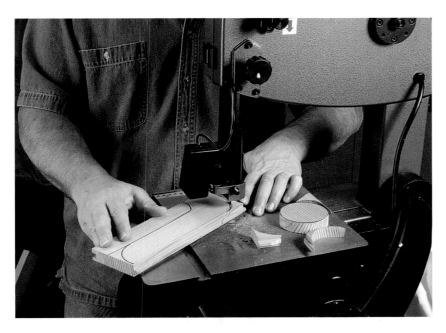

The jig-builder's arsenal of useful hardware includes many different kinds of fasteners, such as nails, screws, bolts, handknobs, hanger screws, and more.

section, I'll discuss the strengths and weaknesses of the most popular and widely available jig-building materials, including natural lumber, plywood, hardboard, MDF (medium-density fiberboard), and plastics—clear, opaque, and slippery. Also included are high-friction coated materials, because they have very useful applications in jig building.

Solid Wood

When lightness and strength are key elements in a jig's design, solid lumber is a valid choice. To minimize warping and dimensional problems, use the most stable lumber you can get. Kiln-dried (KD) vertical-grain fir has straight, uninterrupted grain perfect for making rip fences, stops, and other long, thin jig components. On the downside, vertical-grain fir splinters easily (see the bottom photo at left). Poplar is somewhat softer and less abrasion-resistant than fir, but it's very forgiving to work with, and it's usually cheaper than KD vertical-grain fir. For jig parts that must be strong and wear-resistant, such as cam clamps, dense hardwoods like maple are a much better choice than softwoods.

[TIP] You can save money by using construction-grade lumber (preferably with clear, vertical grain) for making big jig parts. But dry the lumber thoroughly before using it—about a year per inch of thickness.

Plywood

Because of its dimensionally stable cross-ply construction, plywood is free of the kinds of splitting and warping problems common to solid woods. You can also cut strong curved parts out of plywood without concern for

Although cut edges can splinter, vertical-grade Douglas fir is a very stable jig-building material.

grain direction. For this reason, plywood is an ideal material for building jigs. However, with its inconsistent thickness and inferior inner veneers, standard construction-grade plywood isn't the best choice for most jigs. Premium plywoods such as Baltic birch, apple ply, and maple die board are far superior to construction-grade plywoods: They are fabricated with more layers of hardwood veneers (not softwood), and most have thicker, clear-grained face veneers. For the budget-conscious, shop-grade birch plywood is also a very serviceable plywood.

[TIP] To create tight-fitting dado joints for metric-thickness European plywoods or American plywoods (which often deviate ½2 in. or more from their specified thickness), use an adjustable dado set or router bits specifically sized for under-sized plywood.

MDF

Often lumped in a category with particleboard, medium density fiberboard (MDF) is actually a hardboard product. Although comparable in strength and stability to medium-density particleboard, MDF is composed of more finely ground particles, with 10% glue as a binder. MDF is a much better jig-building material than underlayment particleboard because MDF's core is at least 85% as dense as its faces. This makes for clean, smooth edges that are strong and dense enough to hold screws and other fasteners. The dense edges also make it a terrific template material because piloted router bits won't compress and deform MDF edges the way they do particleboard edges.

Because of its exceptionally smooth, dependably flat surface, many woodworkers

Interior plies with knots, or voids like those shown here, weaken construction grades of plywood.

Underlayment-grade particleboard (top) is weak and flaky, compared to medium-density fiberboard (MDF).

prefer MDF over even premium-grade plywoods for making jig bases or jig table surfaces. On the downside, MDF weighs roughly 50% more than plywood, which may be an important consideration if light weight is critical to the setup you're building.

> **WARNING** MDF dust is danger-
> ous for respiratory health, especially for
> woodworkers hypersensitive to the
> formaldehyde in the glue. Use dust col-
> lection and a good dust mask when
> working with this material.

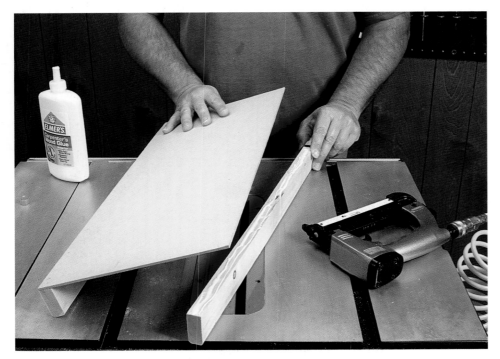

Thin hardboard is a very affordable and practical material to use for making sliding jig bases.

Particleboard

Standard underlayment-grade particleboard —the kind most often sold at lumberyards and home centers—is inexpensive and readily available. Unfortunately, its crumbly inner structure and mediocre strength make it a poor choice for most jig parts. While a poor choice for templates used with piloted bits (the pilot bearing digs into the edge of the template), particleboard is a fine material for veneering cauls and assembly forms where material strength and edge quality aren't crucial.

➤ See "Bending Forms" on p. 230.

Hardboard

With a harder surface than most solid woods and plywoods have, hardboard materials are versatile fiber-based sheet goods useful for making templates and other jig parts. Hardboards come in two types: "screen backed" (smooth on one side with an imprint of the screen pressing process on the other) or "S2S" (smooth and hard on two sides.) Tempered hardboard products such as Masonite are high-density hardboards that have been impregnated with heat-cured resin. Tempering not only improves the board's surface hardness, but it also makes it stronger and water resistant. Masonite's durable edge makes it a favorite for thin, long-lasting templates for pattern routing. Its abrasion-resistant surface also makes it perfect for thin bases for sliding jigs, such as crosscutting sleds, miter cutting carriages, and more.

➤ See "Pattern Routing Templates" on p. 170.

Clear Plastics

Although often lumped under the single heading of "plastic," various plastic materials have radically different working properties and *very* different applications in jig building. Useful by virtue of their transparency, clear plastics are divided into two groups: acrylics and polycarbonates. Common brand names for acrylics include Plexiglas® and Acrylite®. Brands for polycarbonate include Lexan® and Tuffak®. You can distinguish between acrylic and polycarbonate by examining the edge of sheet stock: Acrylic has a yellowish tint to it, while polycarbonate looks dark gray. It's important to know the difference because these plastics have very different properties, making them perfect for some applications and terrible for others.

Acrylic plastics are easy to cut, sand, and even plane and rout. However, because acrylic can shatter when struck, it should never be used for blade or bit guards. Acrylics are also more susceptible to cracking and splitting than polycarbonates, but acrylic sheet stock is stiffer than polycarbonate sheet and doesn't sag or flex as much. This makes acrylic a good choice for clear router table insert plates (see the photo at right). Acrylic parts can be edge glued or face glued using special solvent-type adhesives available from plastics suppliers, or cyanoacrylate "Super Glue" adhesives, which also work very well. Clamp or tape parts into position with joints butted together, wick the thin solvent/glue into the joints, then allow the assembly to dry.

Readily available in thin sheet stock, polycarbonate plastics are less stiff than acrylics

Many kinds of clear and opaque plastic sheet materials are useful for making jig parts.

Because of its stiffness, ¼-in.-thick acrylic plastic is a good choice for router table insert plates.

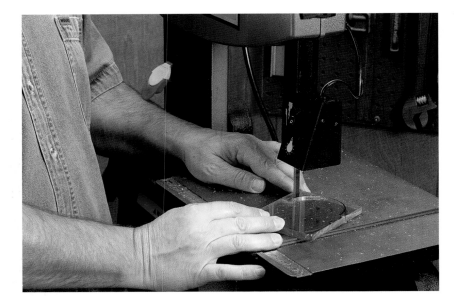

Clear polycarbonate plastics such as Lexan are tough, shatter-proof, and terrific for making jig and machine guards.

Opaque plastics useful for jig building include solid-surface material (rear), ABS (front left), micarta (front right), and phenolic (right rear).

but 30 times more impact resistant, making them virtually shatterproof. Polycarbonate is an ideal plastic for chip deflectors, guards, and other safety devices. You can fasten polycarbonate parts together using machine screws and nuts or by driving threaded fasteners into tapped holes. Alternatively, you can glue parts together with Duco® cement. Because of its thermoplastic nature, thin polycarbonate stock can be heated and bent, allowing you to create curved jig parts and safety guards.

Opaque Plastics

Opaque plastic materials—including phenolic, micarta, ABS, and solid-surface material —are more durable and scratch resistant than clear plastics, but more expensive and harder to find. The most useful of these, phenolic plastic, is twice as expensive as acrylic or polycarbonate, but it is much stiffer and tougher, making it a great choice for production woodworking jigs. Black phenolic plastic is available in $1/2$-in.-thick sheet stock from mail order sources. It is a good choice for machine throat plates as well as sturdy jig bases. You can glue phenolic parts together with epoxy, but for high-strength applications, it's better to drill and bolt parts together.

Slippery Plastics

In the world of jigs and machines, a smoothly gliding workpiece or portable power tool usually results in cleaner, safer cutting or shaping. You can create better sliding jigs by either covering jig parts with plastic laminates or by making parts from slick plastic materials like polyethylene.

▶ INSTALLING GUIDE BARS

You'll get the best performance from any custom sliding jig by carefully fitting its bars and aligning the jig to your machine's miter slots before mounting it. First, trim your shopmade wood or plastic bars or adjust commercial metal bars to fit your machine's individual slots (miter slots aren't always exactly the same width). Apply a little adhesive transfer tape or double-sided tape to the top of each bar and then set the bar atop a few pennies placed in each slot (to raise the bar slightly proud of the surface), as shown in the top photo.

After squaring the edge of the jig's base to the machine table, center the base relative to the blade or cutter, lower it onto the guide bars, and press down so the tape sticks. With wood or UHMW bars, you can simply drill down through the base into the bars and drive screws in place, as shown in the bottom photo. With metal bars, slide the jig out and carefully drill the mounting holes for the bars before installing them with machine screws. Check that the carriage slides smoothly before tightening the screws fully.

Plastic laminates, such as Formica® or Wilsonart®, are easy to bond to plywood or composite substrates like MDF to make smooth, durable surfaces for jig tables and bases. They're also good for covering glue-up fixtures because excess adhesive easily peels or scrapes off for fast cleanup. So-called "ultra-high molecular weight" (UHMW) plastic glide strips are made from dense polyethylene, a tough white material with a very slick surface (see the photo at right). Available in bars and thin sheet stock, UHMW plastic is easily sawn to length or shape. It can be chiseled or planed, but it's impossible to sand. Bars cut to fit a standard

Ultra-high molecular-weight (UHMW) plastic comes as bars, sheet stock, and sticky-backed tape. The latter is good for reducing friction on fence faces.

You can make fences and jig surfaces less slippery by covering them with PSA sandpaper (shown here), or with self-stick, non-slip tape made for stair treads.

miter-gauge slot may be screwed or nailed to jig bases, thus creating very smooth gliding jigs of all kinds. Rolls of thin UHMW material are available in various widths, with a pressure-sensitive adhesive backing. Taping the wide stuff to the face of a jig or fence greatly reduces feed friction.

High-Friction Materials

One way to improve the stability of parts clamped or held in a jig is to apply a high-friction material to contact surfaces to keep the workpiece from slipping or creeping. The easiest way to do this is to stick on a layer of pressure-sensitive adhesive (PSA)

sandpaper, as shown in the photo above. For most applications, choose a paper between 80-grit and 150-grit. You can attach regular sandpaper or other high-friction materials to jig surfaces using a spray adhesive such as 3M®'s No. 77 multipurpose spray glue. To increase the gripping power of clamp faces or fixture surfaces with small surface areas, use a high-friction material such as 3M's SafetyWalk non-skid tape, designed to prevent slipping on stair treads.

Fasteners

A stunning array of fasteners is available to suit the entire spectrum of a jig builder's needs. Your choices range from common nails and screws to hanger bolts and specialized threaded inserts. Simple fasteners—such as production screws and pneumatically

Holding jig parts together calls for a wide range of fasteners, ranging from nails, screws, bolts, and nuts to threaded rod, handscrews, and knobs.

driven nails or staples—allow jig assembly to proceed at warp speed without sacrificing strength or accuracy. Other fastening devices, including T-nuts, studded hand-screws, and taper pins, are ideal for jig parts that must pivot, detach, or articulate for adjustment.

Nails and Staples

Screws or nuts and bolts make very strong connections between parts. But nails, brads, and staples are speedier to use. Whether driven by hand or with an air- or electric-

WARNING *Never* aim a nail gun at any living creature. Also, don't hand-hold parts near the gun's nose, as fasteners can easily deflect in the wood and come out sideways into your hand.

A small compressor is all it takes to power most pneumatic nail guns. Shown here (left to right): a finish nailer, staple gun, and pin gun.

Traditional brass and steel cabinet screws (left) require tapered pilot holes, while modern production and drywall screws (right) don't.

Wood Screws

Wood screws may be more expensive than nails and more time-consuming to drive, but screws have strong advantages. Not only do they create stronger joints between wooden parts, but they are removable. This is most desirable when you're creating a jig from scratch and you're not exactly sure whether it will work as intended. Parts attached with wood screws can be removed and repositioned or replaced as needed.

> **! WARNING** When assembling jigs that will be used near blades and cutters, use brass or bronze screws instead of steel. That way, if a screw is accidentally cut, it won't ruin the blade or spray steel shrapnel.

powered gun, these mechanical fasteners go into place quickly and provide a reasonably strong connection. They not only reinforce glue joints but effectively hold parts in place while the glue dries, eliminating the need for clamps. This is especially handy when you're gluing mitered or beveled parts, which can be hard to clamp. Air-powered nail guns are particularly useful for nailing glued parts together, as a quick pull of the trigger drives a finish nail or brad home before parts have a chance to slip out of position.

[TIP] To keep small hand-driven nails and brads from splitting thin wood parts, blunt the nail's point slightly by tapping it with the hammer before driving it in as usual.

In recent years, most serious woodworkers have eschewed standard zinc-plated wood screws and drywall screws in favor of square-headed production screws. Unlike drywall screws, production screws are made from strong, hardened steel and seldom break. They have deep threads that offer superior holding power, and their square-drive heads rarely strip out, unlike slotted- or Phillips-head screws. Best of all, their cylindrical shanks match the shape of pilot holes made with standard drill bits, whereas traditional wood screws require tapered pilot holes made with a special, expensive tapered drill bit for maximum holding power (see the photo above). Production screws come in steel, brass, bronze, and stainless steel and many head styles, including flat-head, round-head, and washer-head. Washer-head screws are especially handy when you need to attach a thin material like hardboard where regular screw or nail heads might pull through.

Handscrews and Hand Knobs

Standard bolts and machine screws are fine for joining parts but require wrenches or screwdrivers to tighten and loosen. This definitely slows you down when you're making jig adjustments, such as repositioning a stop or relocating a fence or clamping device. With their large, easy-to-grip handles, threaded hand knobs and studded handscrews allow quick setting of jig components without the need for tools (which always seem to go missing just when you need them!).

Hand knobs have a built-in threaded insert and tighten onto the end of a T-bolt, regular screw, or bolt, or onto the machine-threaded portion of a hanger bolt. Handscrews are like bolts with plastic knobs attached to their head end. They can be used with regular nuts or driven into threaded inserts or T-nuts.

► See *"Threaded Inserts"* on p. 27.

Both handscrews and hand knobs come in standard bolt and machine screw sizes ranging from 8-32 size to ³⁄₈ in. A variety of handle styles gives you plenty of choices for your application. The styles include wing nut, three-pronged, knurled (also called rosette), T-shaped, and winged.

Adjustable-handle fasteners have a large, spring-loaded handle that can be easily repositioned on its shaft for applications where there isn't enough clearance for turning a regular knob or handscrew. They also work for securing adjustable fences or stops on jigs where a different handle would obstruct the workpiece, as shown at right.

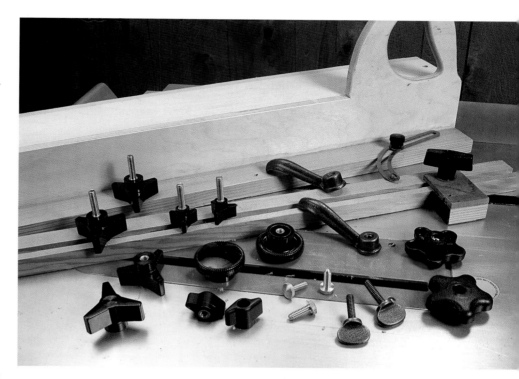

There's practically no limit to the usefulness of plastic-handled hand knobs and threaded handscrews in building all kinds of adjustable jigs.

Lifting the handle of an adjustable-handle fastener allows you to move it out of the way of the workpiece set against the fence of this tilting drill-press jig.

Tapping a taper pin in place creates enough friction to hold stops and jig parts securely without the need for screws or other fasteners.

Taper Pins

Taper pins are commonly used by machinists to accurately locate movable parts like machine fences at square and 45 degrees. A pin fits into a tapered hole that passes through one part and into the other. For the jig builder, taper pins can provide a handy way to accurately locate removable stops, to fix the angle of fences, or to reference the position of auxiliary tables on stationary machines.

Taper pins locate parts more accurately than bolts or wooden dowels, because the pins are self-centering and continue to seat accurately even if the holes become oversized from wear. Furthermore, the friction of a taper pin tapped into parts is actually enough to hold most jig parts together without the need for other fasteners.

Available in many sizes, taper pins come in different lengths and degrees of taper, ranging from #7/0 (almost cylindrical) to #14 (the most taper). Medium-sized pins with a #4 or #5 taper are the most useful for woodworking jig applications. To install a taper pin, first align the parts and clamp them firmly together. Next, drill a hole through both parts that's about the diameter of the small end of the pin. Finally, ream the holes using a special tapered spiral reamer (available from a machinist supply store or catalog), as shown in the photo at left. You can ream using the drill press set to a slow speed, or by hand, with the reamer mounted in a standard tap handle.

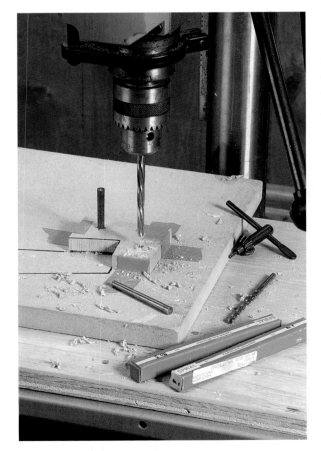

A spiral tapered reamer prepares holes bored with a regular twist bit to receive tapered pins.

Threaded Inserts and T-Nuts

When building a jig with hinged or movable parts, such as a tilting table or adjustable miter fence, it's desirable to attach a bolt or a handscrew into strong metal threads. Threaded inserts and T-nuts (see the photo at right) are designed to be installed in parts made from hard or soft wood, plywood, MDF, particleboard, and even soft plastics. The durable metal threads of these hardware items allow you to refasten handscrews or bolts many times without wearing out the threads. Threaded inserts can be installed into the face or edge of a part and can be used to join parts too thick to easily bolt together. Inserts made of either brass or steel are available in common thread sizes ranging from 6-32 size to ³⁄₈ in.

Threaded inserts and T-nuts add strong metal threads to holes in wood or plastic parts for attaching bolts, handscrews, etc.

► THREADED INSERTS

Threaded inserts have knife-like outer threads made to fit into a hole that's the same diameter as the root diameter of the insert (the body minus the teeth). To install inserts straight and true, use an unplugged drill press. First, cut the head off a bolt that fits the insert you're installing. Chuck the headless end of the bolt into the drill press and then screw on the insert using a pair of jam nuts and a washer to secure it on the bolt. Use a wrench to turn the insert as you lower the drill press's quill (below)

to apply downward pressure. This keeps the insert perfectly straight as it's driven into place. The washer

helps flatten lifted grain as the insert is seated flush to the surface. For jig parts too big for the drill press, use a regular bolt and ratchet wrench to drive the threaded inserts in place (above).

A plastic mallet drives a T-nut into a hole bored in the end of a part cut from 2x4 lumber.

Metal miter gauge slot tracks, extrusions, and guide bars are all very useful for creating a variety of sliding jigs.

T-nuts are somewhat less versatile than threaded inserts, but they're also less expensive and easier to install. T-nuts are commonly available in sizes ranging from 10-24 size to ½ in. After drilling a hole that matches the outside diameter of the T-nut post, you simply pound the T-nut in with a hammer or mallet. Prongs on the flange lock it into the wood to keep it from turning. For greatest holding strength, install the T-nut on the opposite side of the stock so that tightening the bolt pulls the T-nut against the workpiece instead of away from it.

[TIP] Applying a little wax to the threads of a threaded insert will help ease it into the wood. This also helps with regular screws, especially when you're driving them into dense hardwoods.

Tracks and Guide Bars

In addition to fasteners, there is a plethora of other hardware that you can use to quickly build accurate, versatile jigs. For example, extruded-aluminum T-tracks come in many shapes and sizes and offer nearly endless jig-building possibilities. A ready-to-install track can add tremendous versatility to adjustable jigs or clamping fixtures. And commercially made guide bars, as shown in the top right photo, designed to slide in miter gauge slots are easier to install on carriage jigs and slide truer than shopmade wood bars.

Made from extruded aluminum, T-tracks come in different widths, configurations, and slot styles.

Extruded T-Tracks

A T-shaped slot in any wood or plastic jig base can be used to conveniently secure fences, stops, clamps, etc., with T-bolts and fasteners.

You can rout T-slots in a wooden jig, but extruded T-track provides a quicker means of adding a more durable slot to your jig or fixture setup. Made from a strong aluminum alloy, T-tracks come in a wide variety of shapes and sizes to suit lots of different applications.

T-tracks are designed to accept either the head of a standard 1/4-in. hex-head bolt or the flange of a T-bolt or T-nut. In use, the fastener (or T-track accessory) is inserted at the end of the T-track and positioned as desired along the length of the track. When

Some types of T-track are made for use with standard hex-head bolts (rear), while others require use of special T-bolts and nuts.

Used with T-track, flip stops are set relative to a scale and provide an end stop for parts cut to an accurate length.

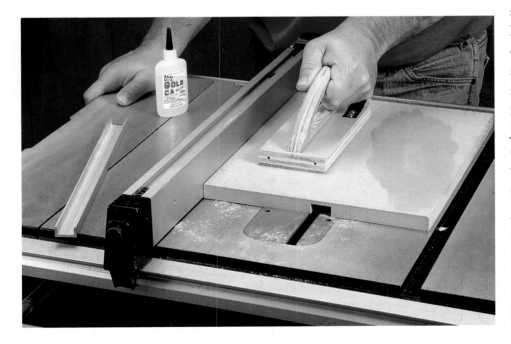

Most miter slot track extrusions are sized to fit into a 1-in.-wide slot, cut in multiple passes on a tablesaw fitted with a dado blade.

the fastener is tightened, the track prevents the bolt head, stud, or nut in the track from turning, allowing quick one-handed tightening and loosening.

The most basic type of track for simple jig setups has only a single T slot. The track is installed with either screws or glue into a simple dado or groove. Single-slot T-tracks are unbelievably handy for creating versatile hold-down tables as well as clamping and assembly fixtures.

Wider T-tracks with two or more T-slots are designed to serve several different jigging purposes. You can use L-shaped T-track to create a fence for a cutoff saw, router table, or other machine as seen in the photo at left.

Even wider extruded tracks, with four or more slots, are useful for making high fences, clamping jigs, and other devices. In addition to standard T-bolts, T-nuts, or hex-head bolts, most tracks accept a wide range of accessories, such as flip stops, micro-adjusters, and cutter guards. Flip stops are intended to work with tracks mounted to fences on miter gauge cutoff jigs, mortising jigs, etc. The beauty of flip stops is that one or more stops can be flipped out of the way when necessary without losing their position on the fence.

[TIP] Make sure to check whether your desired T-track accommodates regular hex-head bolts or T-bolts and nuts. Most brands and styles of track accept only a particular size and style of fastener.

Miter Slot Track

Miter slot track is a handy product for adding a durable slot to a jig base or table to accommodate a regular miter gauge or any sliding sled or carriage. Extruded aluminum miter slot track can be inlaid into solid wood, plywood, MDF, or nearly any material that's at least ³⁄₄-in. thick. Made by a number of different manufacturers, tracks come in 24- and 48-in. lengths. Most brands are made to fit a 1-in.-wide, ¹⁄₂-in.-deep routed or sawn groove. Some brands of track include pre-drilled holes for screwing the track to the bottom of the groove. Others are designed to be glued into the slot using epoxy, polyurethane glue, or a thick-viscosity cyanoacrylate (CA) glue.

Miter Guide Bars

The trickiest part of building almost any miter-slot guided carriage is making and fitting the guide bars that run in the machine's miter gauge slots. Wooden bars, even when perfectly sized and aligned, tend to run loose in the dry winter months and tight during

► See *"Slippery Plastics"* on p. 20.

humid summers. I've had much better success fitting my shopmade carriages with bars made from UHMW polyethylene or commercially made metal guide bars.

Both plastic and metal bars are straight and stable and will glide smoothly in any standard ³⁄₄-in.-wide, ³⁄₈-in.-deep miter gauge slot or miter slot track. Polyethylene bars come ready-made, or you can make them yourself from UHMW sheet stock. If

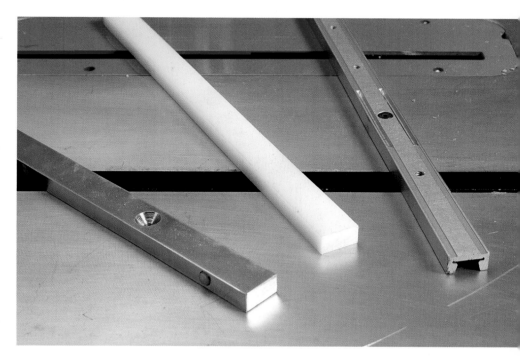

Commercially available miter bar stock includes (left to right) steel with adjustable plastic inserts, UHMW plastic, and Incra® aluminum bar.

you make your own, saw the bar oversize first, then plane its width a little at a time until it fits its slot perfectly.

Other Jig Hardware

Sometimes a custom jig calls for a specific piece of hardware to make it function as desired. For instance, you may need to add a measuring tape to a fence jig to ensure accuracy. Or you may want to use powerful rare-earth magnets to temporarily attach a large hold-down assembly to your cast-iron tablesaw table. Other hardware items such as air-powered cam clamps, dust ports, and vacuum devices are discussed in the sections where they are implemented.

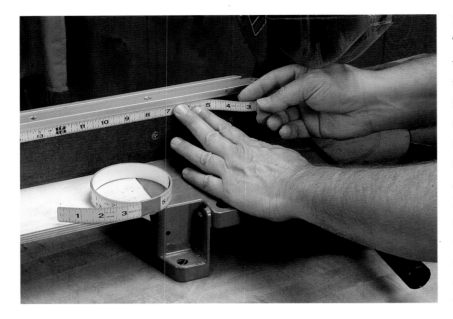

Self-stick measuring tape makes it very easy to add a scale to a crosscut fence or any jig with adjustable parts.

Powerful rare-earth magnets are strong enough to temporarily attach fences, hold-downs, and other jigs to steel or cast-iron tool surfaces.

Self-Stick Measuring Tape

To get accurate, repeatable cuts using fences with T slots and sliding stops, attach a measuring tape to the fence or to the T-track itself. A self-stick tape measure has a pressure-sensitive, peel-and-stick backing for easy attachment to wood fences, aluminum extrusions, or any other smooth, clean surface. Available with scales that read either right to left or left to right, self-stick tapes provide a convenient way to add an adjustment scale to a fence, part-positioning fixture, or adjustable jig component. For even more accurate setting of stops to the tape, a cursor with a fine crosshair may be fitted.

▶ See *"Flips Stops"* on p. 192.

Rare-Earth Magnets

"Rare-earth magnets" (neodymium magnets) are made from a combination of neodymium, iron, and boron (NdFeB), creating one of the world's most powerful permanent magnetic materials. Available in several diameters, these small disc-shaped magnets have tremendous holding power when placed against steel or iron parts. They're perfect for temporarily attaching a guard to any cast-iron machine table. Similarly, four $3/4$-in.-dia. magnets set into the base of a wheeled hold-down will firmly "clamp" it to a bandsaw table. These magnets will even hold a fence jig securely enough for resawing operations.

> ⚠ **WARNING   Rare-earth magnets are so strong that they can wreak havoc with electronic devices and magnetic media. Never place these powerful magnets near computers, PDAs, watches, cassette tapes, or credit cards.**

Tools and Techniques

OST OF THE TOOLS REQUIRED to build jigs are the same as those used to build furniture and cabinetry. And most practices for building solid, durable jigs from wood and wood composite materials are the same as for building quality furniture. However, in many cases jigs must be built to higher tolerances in order to perform precise, repeatable operations. Furthermore, they must be durable enough to withstand the forces imposed upon them by powerful machines. In this section, we'll explore joinery, gluing, finishing, and other woodworking techniques that apply to jig building. Also discussed are construction methods for other materials, such as tapping metal and plastic parts to accept bolts and screws.

Jig-Building Tools

In addition to the usual woodworking tools that are essential for building jigs from wood, composite wood materials, plastics, and aluminum, there are several tools we can borrow from the metal machinist's tool chest. These items, which include dividers, rules, dial indicators, and 1-2-3 blocks, are very helpful for aiding accuracy.

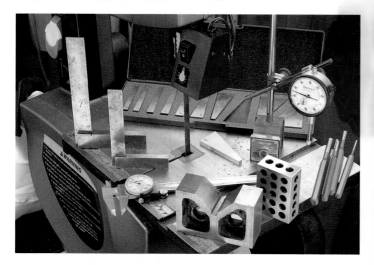

Machinist's tools—including calipers, dial indicators, and 1-2-3 blocks—all help to ensure accuracy when you're building and setting up jigs.

Measuring calipers are useful for checking both inside and outside dimensions of parts, as well as the depth of slots and holes.

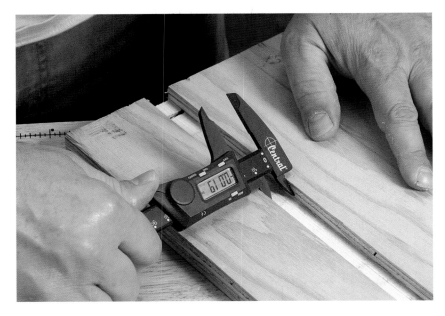

Digital calipers feature a liquid crystal display readout that can convert between metric and imperial measurements.

Measuring Calipers

Measuring calipers are extremely handy, simple-to-use tools that gauge very accurate measurements. Most calipers offer three ways to check a measurement: The outside-measuring jaws will check part lengths and thicknesses; the inside-measuring jaws will check slot width or hole diameters; and a depth rod will gauge the depth of holes and grooves. Modern measuring calipers come in three popular styles: Vernier calipers, which read measurements in thousandths of an inch and micrometers from a special Vernier scale shown on the beam; dial calipers that show measurements (thousandths, millimeters, and/or fractions of an inch) on a dial similar to a watch face; and digital calipers that display dimensions on a small liquid crystal display (LCD) screen, as shown at left. Push a button and most models instantly convert between imperial, decimal, or metric measurements.

Engineer's squares are useful for checking the accuracy of machined wood parts and for adjusting blades and cutters on shop machines.

Engineer's Squares

Made in several sizes, engineer's squares are very sturdy and accurate try squares that are ideally suited to the precise layout and assembly tasks necessary for building quality jigs. An engineer's square consists of an unmarked, heavy steel blade rigidly mounted to a thick solid-steel body. The smaller-sized 4-in. square is excellent for layout work on small jig parts, while the big 12-in. square helps you align parts properly when constructing large assemblies. An 8-in. square is good for general duty, as well as for setting blades on tablesaws, aligning drill press tables, etc.

Machinist's Blocks

V-blocks and 1-2-3 blocks are mainstays in any machine shop. Each is useful for a wide range of machine set-up purposes and parts-machining operations. Made of hardened steel with squarely ground sides, a 1-2-3 block is a rectangular piece of metal 1 in. thick, 2 in. wide, and 3 in. long. Most are covered with holes—some of them threaded—to save weight and to provide a variety of mounting options. Because their edges are precisely square, 1-2-3 blocks are perfect for spacing and positioning parts that are being drilled or machined, as well as for keeping parts square as they are fastened together. For example, use one or two of these heavy blocks to hold a wood fence atop a jig base plate as it's glued and screwed into place, as shown in the top photo at right.

A V-block is a heavy, rectangular steel block with a V-shaped hollow on one edge. The V is perfect for supporting either round or square-edged stock during drilling or machining operations, as shown in the bottom photo at right. Once the bottom of the V is accurately positioned under the centerline of a drill press quill, holes will always be centered, regardless of the size of the bit you use or the stock you drill.

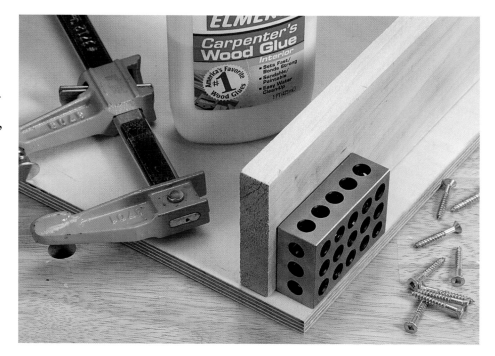

A machinist's 1-2-3 block helps keep wood parts aligned during glue-up, ensuring precise jig construction.

A heavy steel V-block holds dowels and other cylindrical work-pieces steady during drilling or machining operations.

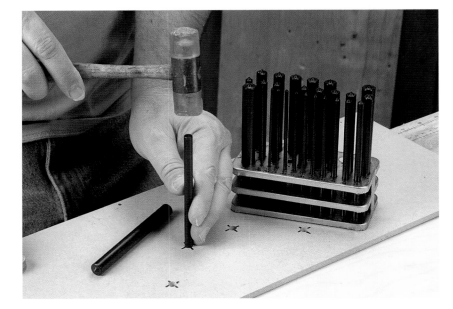

A transfer punch fits through a hole in a layout template to mark the precise location of the hole on the workpiece underneath.

A basic butt joint requires reinforcement with glue and nails or screws for strong assembly.

Transfer Punches

Whenever you need to accurately transfer the location of an existing hole or routed slot to another part, a transfer punch is the tool to reach for. Available in sets of different diameters, transfer punches are metal rods with small raised points in the center of one end. This machinist's layout tool is very easy to use when you're working with metal, wood, and other sturdy materials. With the part in position over the work surface, choose a punch that matches the diameter of the hole and then slip it in and give it a tap. The punch leaves a small depression in the lower part (see the top photo at left) that will center a drill bit for boring a hole in that exact location. Transfer punches provide dead-on accuracy when you're mounting large pieces of hardware, attaching stops, fences, and other jig parts, or when you're locating threaded inserts that are perfectly aligned with routed slots, say, in the compass of an adjustable table jig.

Joinery

While throwing a jig together with nails and glue may be acceptable, there's no substitute for using sound wood joinery to attach parts. In lieu of the strongest traditional joints (mortise and tenons, dovetails, etc.), simpler butt joints, biscuit joints, rabbets, and dadoes are adequate for most jig-building projects. For making flat, rigid jig tables or other surfaces, torsion box construction is an excellent choice.

Butt Joints

If you're building jigs in a hurry, then gluing and nailing together butt-joined parts is usually the way to go. A butt joint is one where two parts with square-cut ends are fastened together, as shown in the bottom photo on the facing page. Butt joints are very fast and easy to cut and adequate for many jig box and frame constructions. However, you must reinforce butt joints with nails, staples, wood screws, splines, or biscuits.

To help align a butt joint during assembly, use nails and screws in concert: After applying glue to both mating surfaces, carefully position the parts and drive a nail in at each end of the joint to keep parts from sliding out of alignment (see the top photo at right). Then drive the screws.

Biscuit Joints

Plate joinery biscuits are a quick and handy way to reinforce a basic butt joint. Biscuits installed in slots cut by a plate joiner machine are ideal for creating strong joints in solid wood, plywood, or wood composite sheet goods. A row of biscuits added to the corners of a butt-joined frame or to the intersection of two panels offers tremendous strength and facilitates alignment and clamping during glue-up. Biscuits can also be used to reinforce end-to-end panel connections, as well as mitered frame and box joints.

On wide parts, biscuit slots should be spaced every few inches along the joint. After plunge-cutting the slots in both parts (see the bottom photo at right),

A few pneumatic nails quickly driven into these glued butt joints keep parts from sliding out of alignment as screws are driven to strengthen the joint.

To join two panels at right angles, plunge biscuit slots into the face of one panel and into the end of the mating panel.

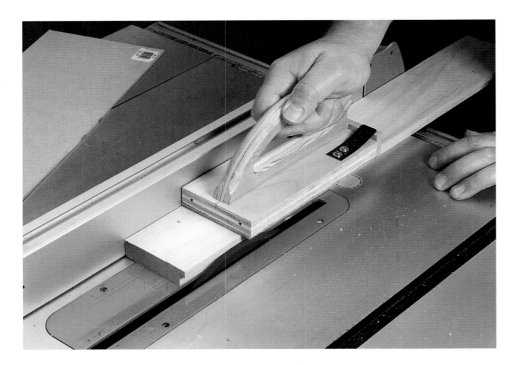

A rabbet can be cut on the edge of a workpiece in two passes on the tablesaw using a regular sawblade.

Special "undersized" bits are made for routing snug-fitting dadoes and grooves for commercial plywoods.

apply glue to the slots and biscuits, assemble the joint, and clamp it up. For mitered joints and assemblies that are difficult to clamp, you can drive a few screws through the parts to hold the joint together until the glue dries.

Rabbets and Dadoes

Rabbet and dado joints are not only quick and easy to cut, but they add connective strength to joints and make jig parts easier to align and assemble. The joints work well in practically all solid wood and composite wood products as well as thicker plastic stock. A rabbet is a stepped joint used to join the corners of frames and box-like constructions. A ledge is cut into the end of one part with either a dado blade on the tablesaw or a rabbet bit in the router (see the top photo at left). The width of the rabbet should equal the thickness of the part it will accommodate. Rabbet depth can vary but shouldn't exceed two-thirds of the stock thickness.

A dado is a square-edged channel that accepts a part joined at a 90-degree angle. For a nice, tight joint, the width of the dado should exactly equal the thickness of the mating part, whose square end fits into the dado. The depth of a dado shouldn't exceed half the thickness of the stock it is cut into. As with a rabbet, dadoes are cut either with a dado blade on the tablesaw or with a router, after which they are glued and nailed or screwed in place.

Routing Dadoes in Plywood

Getting tight-fitting dado joints can be difficult when you're working with typically sized plywoods—¹⁄₄ in., ¹⁄₂ in., ³⁄₄ in., etc.—since they are not exactly as thick as stated. However, you can cut dadoes to a precise width with an adjustable stack dado on a tablesaw or with special "undersized" bits made for the purpose (or metrically sized bits, for European plywoods), as shown in the bottom photo on the facing page. For cleaner cuts with less tearout, select bits with a "down shear" design, which puts a slight downward pressure on the surface of the plywood during cutting. Another way to rout odd-dimension dadoes is to use an adjustable router edge guide (see the top photo at right) and a straight router bit that's smaller in diameter than the width of the required dado. Take one cut and then readjust the edge guide to widen the first pass to the exact desired dado width.

Routing T-Tracks

Solid, permanent joints are in order for durable, long-lasting jigs. But adjustable jig components, including stops, fences, and fixture clamps, must be removable and repositionable. T-tracks and the special fasteners that slide into them provide a very strong and easy-to-use system for mounting a wide variety of jig components. You can buy T-track as a ready-made extrusion that installs in a simple groove, or you can cut a T-slot using a router, as shown in the bottom photo at right. Although routing your own T-slot into plywood, MDF, or even particleboard won't provide as strong a mount as an extruded metal track, it will be very serviceable.

► See "Extruded T-Tracks" on p. 29.

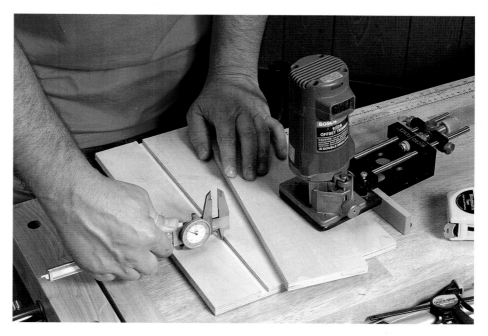

Cutting odd-sized dadoes and grooves is easier with a premium router fence that allows for fine adjustments.

Routing a T-slot is a two-part process: First, rout or saw a standard groove and then reshape it using a special T-slot router bit.

TORSION BOX CONSTRUCTION

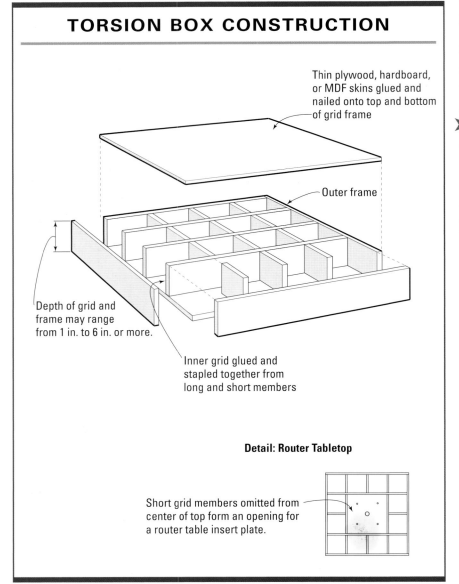

Thin plywood, hardboard, or MDF skins glued and nailed onto top and bottom of grid frame

Outer frame

Depth of grid and frame may range from 1 in. to 6 in. or more.

Inner grid glued and stapled together from long and short members

Detail: Router Tabletop

Short grid members omitted from center of top form an opening for a router table insert plate.

The first step in routing a T-track is to create a ⁹⁄₃₂-in.-wide groove, cut with either a dado blade on the tablesaw or with a straight bit in the router. Once this groove is cut, the bottom of the groove is widened using a special T-shaped bit. This slot will accept either ¼-in. T-bolts or T-nuts as well as standard plumbing toilet bolts (used to attach a toilet to its flange).

Building a Torsion Box

Torsion-box construction provides a terrific way of creating very strong jig bases, router tabletops, clamping and fixture tables, and more.

▶ See *"Air Table"* on p. 130.

A torsion box has an outer frame and an inner grid sandwiched between a thin top and bottom skin (see the drawing at left). It's built to be strong and light like an airplane wing. You can make the outer skins from plywood, MDF, or hardboard as thin as ⅛ in. The frame and inner grid can be made from just about anything, including plywood, particleboard, or No. 2 pine. The combined thickness of the frame and grid can be as thin as 1 in. or can exceed 6 in. The thinner the skin, the closer the grid spacing needs to be to keep the top rigid: For ⅛-in.-thick skins, use a 2-in. grid spacing; a 4-in. to 6-in. spacing is fine for ⅜-in.-thick skins.

Amazingly, minimal joinery is necessary to connect the parts of the grid to each other. The short components are just glued and stapled to the long ones to hold their edges in a plane until the skins are attached (see the top photo on the facing page). Assemble the box on a dead-flat bench or machine top after spreading glue liberally on the edge of every grid member. Attach the skins using brads or narrow-crown staples every few inches. For a durable surface, cover the top skin with plastic laminate.

If you're making a router tabletop, you can frame out an open section in the middle of the torsion box. (See the detail drawing on the facing page.) Apply the skins and then cut them out to accommodate the router section. A lip routed in the top opening supports an insert plate that mounts to the router.

Gluing Up

Vibration from machine tools can make fasteners lose their hold over time, jeopardizing the accuracy of a jig. Therefore, fixed jig parts such as non-adjustable stop blocks and fences that must maintain alignment should be fastened with glue in addition to screws or nails. While standard yellow carpenter's glue is just fine for most wood-to-wood gluing jobs, epoxy or cyanoacrylate glues also have some very useful applications in jig building.

[TIP] **Whatever glue you use, never glue solid wood parts wider than 4 in. cross-grain to each other, or you risk future joint failure due to expansion/ contraction problems.**

Epoxy

Two-part epoxy glue offers terrific versatility for jig building, because it bonds to a wide assortment of materials, including wood, composite materials, metal, and many plastics. And it will cross-bond dissimilar materials, such as metal or plastic to wood. Epoxy also has excellent gap-filling properties, so it's useful for gluing joints that don't fit snugly together.

There are a few things to remember when choosing and using epoxies: First, the 5-minute type isn't as strong in the long run as the type that takes a full 24 hours to cure. Secondly, always dispense

The MDF parts of this torsion box grid need only be joined with glue and staples driven into the edges of adjacent parts.

With the torsion box assembly lying on a dead-flat surface, glue and nail the thin skins onto the top and bottom of the gridwork.

High bonding strength and the ability to glue dissimilar materials together make two-part epoxy adhesives very versatile adhesives for building jigs.

Commonly called "super glue," cyanoacrylate adhesives form strong bonds and dry almost instantly when sprayed with an accelerant.

epoxy's two parts equally, and mix them very thoroughly for at least 30 seconds before applying. Third, for the best bond, surfaces must be clean, and wood should be freshly machined. Finally, let parts sit after clamping, leaving them alone for at least as long as the recommended set time of the epoxy.

[TIP] Don't discard your epoxy-mixing pallet or container after glue-up; the remaining glue on it will inform you when the joint has cured.

Cyanoacrylate Glue

Commonly called "super glue" or "CA glue," cyanoacrylate adhesive is known for its great strength and ability to bond many different materials. CA glues are available in different viscosities: The thin variety works better for smooth, non-porous materials such as metal and plastic, while the thick type works best for porous woods. When sprayed with accelerator, CA will set instantly, making it very useful for attaching small parts that are difficult to clamp, or for bonding thin parts clamped atop wax paper.

See *"Mortise Template"* on p. 54.

CA is also great for holding hardware in alignment while screws are installed. To instantly bond parts together, apply the CA to one part and spray accelerator (sometimes called "kicker") on the other before pressing or clamping the parts together for a few seconds while the glue sets, as shown in the bottom photo at left.

Finishing

As with any other woodworking project, finishing is the last step in creating stable and durable jigs and fixtures. A quick coat of finish on raw wood, plywood, and even MDF and hardboard serves to protect the wood from moisture and to reduce the effects of humidity, which might otherwise split solid parts or warp flat panels. A sturdy finish can also improve a jig's wear resistance and make it easier to clean.

In contrast to finishing furniture or cabinets made from fine hardwoods, the task of finishing jigs and fixtures is considerably less demanding. Wear-resistant finishes are best, such as brush-on or spray-on varnishes and polyurethanes. My favorite jig finishes are wipe-on polyurethanes, which have excellent durability and which apply easily and dry to the touch in 15 minutes. You apply a thin layer on the wood using a clean rag or pad, let it soak in a bit, and then wipe off the excess. For best results, apply two coats (especially to open-grained woods like oak), waiting the recommended dry-time between coats.

[TIP] On jigs made from wood parts that have only been lightly sanded, applying a wipe-on finish with a nylon abrasive pad such as "Scotch-Brite®," will help smooth edges and prevent splinters.

Wipe-on polyurethane finishes are very easy to apply and help protect wood jig parts from wear, dirt, and the effects of humidity changes.

▶ RECORDING JIG INSTRUCTIONS

Some jigs require special machine setups that can be difficult to remember over time. To save head-scratching and searching through old notes every time you use the jig, it's wise to record the pertinent setup and use information directly on the jig as shown at right. This might include the type of blade or bit used, its height or depth setting, the position of fences or stops, and

the order of operations. Note the details using a fine-point permanent marker and then protect the information from wearing off by coating the jig with a finish.

Marking and Setup Jigs, page 46

Fences and Guides, page 63

Sliding and Pivoting Jigs, page 91

Jig Tables, page 121

Jigs for Portable Tools, page 139

Templates, page 167

Woodworking Jigs

ROM THE FIRST PENCIL LINE marking the length of a part to the last sanding and smoothing operation, jigs are important tools that make building your wood projects faster and easier. There's an endless variety of jigs you can make to suit the specific needs of an operation using stationary woodworking machinery, portable power tools, or hand tools.

Marking jigs help us lay out precise dimensions, joinery, and consistent shapes, while setup jigs are invaluable for adjusting machinery. Fences, guides, and sliding jigs all add consistency and safety to our machining operations by aligning and guiding workpieces for accurate cutting, shaping, boring, planing, and sanding. Tabletop jigs can extend the function of many shop machines, while portable power tool jigs increase control and accuracy and extend the range of capabilities of routers, circular saws, jigsaws, and more. Finally, templates and patterns help us create shapely parts as well as precise, tight-fitting joinery.

Marking and Setup Jigs

Marking Gauges

➤ Dedicated Marking Gauges (p. 48)
➤ Center Marking (p. 50)

Layout Jigs

➤ Beam Compass (p. 52)
➤ Ellipse Jig (p. 53)

Templates

➤ Mortise Template (p. 54)
➤ Angle and Dovetail Templates (p. 55)
➤ Corner and Part Templates (p. 56)

Machine Setup

➤ Story Sticks (p. 57)
➤ Depth Gauges (p. 58)
➤ Setup Gauges (p. 59)
➤ Angle Gauges (p. 60)
➤ Router Cradle (p. 61)

THERE ARE LOTS OF COMMERCIALLY available devices to help us mark lines and circles on our parts, many of which are adjustable to accommodate different circumstances. However, these devices take time to set and may slip out of adjustment. The fact is, we don't always need a marking device to be adjustable, because we very often mark our stock and set our machines to common dimensions or settings. For example, how often have you taken the time to set an adjustable marking gauge to scribe a line $\frac{1}{2}$ in. or $\frac{3}{4}$ in. from an edge?

In this section, I'll show you how to make dedicated gauges that will quickly take care of many everyday marking and layout tasks. Some of these shopmade gauges will help you mark cutlines, locate screw holes, and align hardware

to an edge. Other handy devices here include jigs for finding centers and for drawing curved lines, circles, and arcs. You'll see how to make a large beam compass and an amazingly simple jig for laying out ovals and ellipses. You'll even find out how to use your fingers as a simple marking device.

➤ See *"Quick Finger Gauge"* on p. 47.

I'll also discuss templates as layout tools. While you might think templates are primarily for pattern-cutting with a router (see p. 159), they're also useful for a variety of marking and layout tasks. Simple curve and radius templates help lay out identical cutlines quickly and easily. Angled templates are

great for marking miter cuts, while mortise templates and dovetail template jigs are useful for laying out joinery. You can even use a template jig such as the one shown in the center photo below to transfer complex patterns onto multiple identical parts for attaching screws and hardware.

▶ See *"Pattern Routing Jig"* on p. 159.

Custom setup jigs are also handy devices for setting adjustments on shop machines. For example, setup gauges can help you quickly set the height of router bits or shaper cutters without a ruler or calipers. A triangular setup gauge (right photo below) is useful for accurately setting miter and/or bevel angles on crosscut saws and tablesaws. Even a plain skinny strip of wood can serve as a "story stick" to record a project's every dimension and to make machine setup a cinch. At the end of this section, you'll find one of my favorite setup jigs: a router cradle that holds the tool while locking its arbor shaft, for painless bit changes.

▶ QUICK FINGER GAUGE

Who says you have to use a marking gauge or jig for every layout job? When you need to draw a line parallel to the edge of a workpiece to mark for a row of screw holes or a piece of hardware, try this time-saving trick: Grasp a sharp pencil tightly with your finger, then press your fingertips against the edge of the workpiece. Carefully extend the pencil so that the point contacts the wood at the desired marking distance. Now draw the pencil along the workpiece, keeping your fingertips pressed tightly against the edge, as shown above. With a little practice, you'll be surprised at how precise and straight a line you can mark this way.

Shopmade marking gauges are easy to make and can be customized to suit a variety of project layout tasks.

Use a thin hardboard template and a transfer punch to accurately mark the position of holes for driving fasteners and mounting hardware in multiple identical workpieces.

Hardboard triangles cut to various precise angles serve as quick, accurate setup gauges for a miter saw.

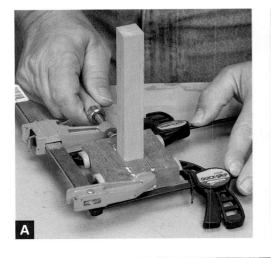

Dedicated Marking Gauges

Unlike adjustable marking gauges that take time to set and that can accidentally come out of adjustment, a dedicated marking gauge is fixed and can mark dimensions with great accuracy and reliability. The gauge shown here consists of a fixed fence encircling a beam. It can mark eight common dimensions, although you could design a gauge to mark any variety of dimensions you wish. You can use it to mark more dimensions by placing a shim between the fence and work-piece as you mark (see photo E). To make the gauge, first cut the ¾-in.-square by 6¾-in.-long beam from straight-grained hardwood. Make the fence from ¾-in.-thick hardwood, cutting two pieces to 2¾ in. by 1 in. and two pieces to ¾ in. by 1 in. Glue the pieces up around the beam **(A)**, but take care not to glue them to the beam.

After the glue cures, position the fence so that one face is exactly 1½ in. from one end of the beam and mark its position. Then, measuring from that same fence face outward, make a mark at ¼ in., ½ in., ¾ in. and 1 in., allocating one mark only to each face of the beam. On the opposite end of the beam, place marks at 1½ in., 2 in., 3 in., and 4 in. away from the face of the fence.

Slide the fence off the beam and use a small drill or a rotary tool to drill a tiny pilot hole at each mark to accept a no. 18 brad **(B)**. Then slide the fence back on the beam and reposition it to its mark. After drilling a countersunk screw hole through the edge of the fence, secure it to the beam with a screw. Drive a brad into each tiny pilot hole and then clip its head off about ¹⁄₁₆ in. above the beam surface with a pair of diagonal cutters **(C)**. Snip with the cutters oriented perpendicular to the beam so the cut end will form

a wedge-shaped tip that's parallel to the fence. If a brad needs slight repositioning, you can fine-tune it by bending it or by filing its tip toward the desired direction with a needle file **(D)**.

You can also make dedicated marking gauges for particular layout tasks by mounting a beam of the desired length to a short fence. For best alignment and strength, glue the beam into a dado cut in the center of the fence piece **(F)**.

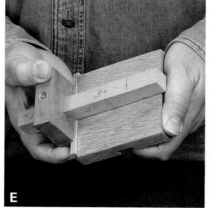

Pencil-Marking Gauges

If you prefer marking your work with a pencil line rather than a scored line, you can make a pencil-marking gauge. Instead of driving a small nail into a gauge's beam, drill a hole that's slightly smaller in diameter than a round-shank pencil **(A)**. Locate the hole so that the distance between its center and the fence equals the desired marking distance. On the bandsaw, cut down the center of the beam about an inch past the hole **(B)**. This kerf allows the hole to spread slightly when the pencil is inserted, but still allows enough clamping force to hold the pencil firmly in place.

► See *"Quick Finger Gauge"* on p. 47.

[TIP] To ensure the accuracy of any pencil gauge, make sure the pencil's point is sharp and concentric to the shank.

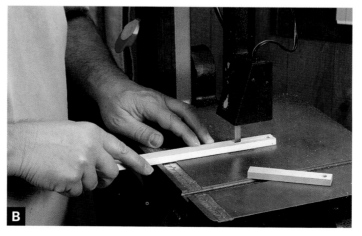

Center Marking

A center-finding jig that allows you to quickly mark the center of square, rectangular, round, or polygonal stock is a must-have jig for woodturners. Make the base of the jig from two pieces of 6 in. x 12 in. x ½-in.-thick plywood or MDF. Glue and nail the parts to form an L-shape, using a pair of machinist's 1-2-3 blocks to keep them square **(A)**.

► See *"Machinist's Blocks"* on p. 35.

Screw a 4-in. plastic 45-degree triangle (available from art supply stores) to one end of the base with several small wood screws and washers **(B)**. Make sure to align the hypotenuse with the inside corner of the base. To use the jig, press the workpiece firmly into the corner and against the plastic square, and draw a diagonal line across the work **(C)**. Rotate the work 90 degrees, and mark it again. The center point is where the two lines cross.

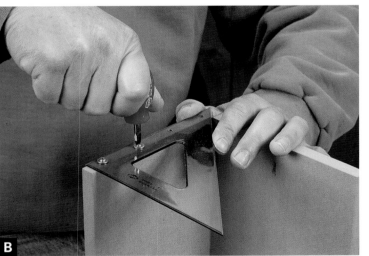

To make a versatile center-marking jig that'll work with stock ranging from ½ in. to 13½ in. wide, cut a 16-in.-long, 1½-in.-square bar from straight-grained stock. Bore a centered hole just big enough for a pencil through each long face of the bar. Then drill a pair of ¼-in.-dia. holes, 1 in. deep on each of the bar's four long faces, each pair equidistant from the center and spaced 2 in., 6 in., 9 in., and 14 in. apart, respectively **(D)**. Apply glue and drive a 2-in.-long dowel into each hole **(E)**. To use the jig, choose the side with the smallest spacing that will span the workpiece, insert a pencil, and slide the jig along the work, keeping the opposing dowels in firm contact with the edges **(F)**.

A

B

C

D

E

F

Beam Compass

A beam compass will draw large circles and arcs for laying out tabletops, arched moldings, etc. The beam of this compass—designed to draw curves up to a 6-ft. radius—is made from two strips of ½-in. x 1½-in. doorstop molding clamped together with a pair of large notebook clips **(A)**.

To make the compass, cut the pivot strip 42 in. long, bevel one end to 45 degrees, and color it with a marker pen to serve as a cursor. Glue a 3-in. square of ½-in.-thick plywood or MDF to the other end for a base. Using a drill bit whose diameter matches that of an 8d nail, drill a small hole vertically through strip and base to accommodate the jig's pivot point **(B)**. Cut an 8d nail to 2 in. long with a hacksaw and then resharpen its point with a file **(C)**. Carefully drive the nail into the hole.

Cut the mating strip to 48 in. long and cut another piece 3 in. long for the pencil clamp. After sawing a small V-groove near the end of each piece to accommodate a pencil **(D)**, screw the clamp and pencil to the strip **(E)**.

A self-stick measuring tape on the inner face of the pencil strip allows you to quickly set the compass to an accurate radius. You'll need the 48-in.-to-72-in. section of a tape that reads right to left. Clip the beam strips together, set the distance between pencil and pivot points at exactly 48 in., and mark a line where the beveled cursor meets the other strip. Align the self-stick tape with this mark **(F)**, and you're ready to go.

[TIP] **Save the leftover sections of self-stick measuring tapes for use on various jigs.**

Ellipse Jig

One way to draw an accurate elliptical shape for an oval picture frame, plaque, or tabletop is to use a commercially made ellipse jig **(A)**. In addition to its use in marking, such a jig can also be used to cut out elliptical shapes with a router. But if you don't have this jig handy, here's another method that'll get the job done. All you need is a carpenter's framing square and a jig stick. First, cut the workpiece into a rectangle that's exactly as wide and long as the desired ellipse, and draw a horizontal and vertical line through the exact center **(B)**. The lines divide the rectangle into four quadrants: Measure and record the length and width of one quadrant.

Make the jig stick from a piece of ¾-in. by ½-in. hardwood that's about two-thirds the total length of the desired ellipse. Drill and slit one end of the stick to hold a pencil, as described on p. 49. Then drive two small finish nails into the stick to serve as guides that will run against the edges of the square when you're drawing the ellipse **(C)**. Space one nail the same distance from the pencil point as the length of the quadrant. The spacing for the other nail should match the width of the quadrant.

▶ See *"Pencil–Marking Gauges"* on p. 49.

The ellipse is sketched in four stages. In each stage, the framing square is aligned with the lines defining the quadrant below the one you're drawing **(D)**. Starting with the pencil at the long end of the workpiece, pivot the jig stick while keeping each of the nail guides firmly against the edges of the square, thus drawing a quarter of the ellipse in a single, fluid motion. Repeat the process in the other quadrants to complete the ellipse.

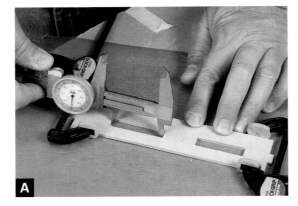

A

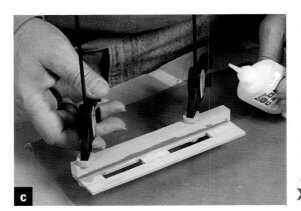

B

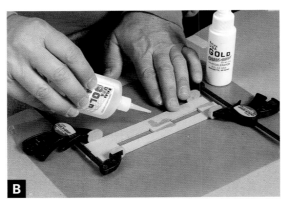

C

D

Mortise Template

Mortise marking templates are a great aid for accurately laying out mortise-and-tenon joints, as well as for marking hinge mortises or hardware recesses. Made of ⅛-in.- or ¼-in.-thick hardboard, a mortise template has an opening that's the same shape and size as the desired mortise, providing you with a guide for your pencil or layout knife.

It's easy and accurate to assemble the template from several strips of hardboard cut to the proper dimensions and glued together. You can make a template any size you like. For this example, I've made one that suits two ⅜-in.-wide mortises of different lengths that I commonly use for 2-in.-wide frame members.

Start by sawing three strips of hardboard: one ⅜ in. wide, one ⅝ in. wide, and one ¾ in. wide. Cut a 1-in. length off the ⅜-in. strip and then crosscut the remaining piece in half. Now, working on a sheet of waxed paper, assemble the five pieces of hardboard together, sandwiching the ⅜-in.-wide strips between the ¾-in. and ⅝-in. strips, with the 1-in.-long strip in the middle **(A)**. Position the ⅜-in.-wide strips to create a 1¾-in.-long opening and a 2-in.-long opening. Lightly clamp the assembly together and check the dimension of the template openings, tapping the ends as necessary to tweak the size. Glue the parts together by applying a few drops of thin cyanoacrylate glue to each seam **(B)**.

▶ See "*Cyanoacrylate Glue*" on p. 42.

After trimming the ends of the template square, glue a strip of MDF onto the template to serve as a fence **(C)**. In use, the fence registers against the workpiece, automatically centering the template on ¾-in.-thick stock **(D)**. Once this is positioned, you can then mark the outline of the mortise with a striking knife or fine pencil.

Angle and Dovetail Templates

It's easy to consistently and accurately mark any angle on a workpiece using a simple marking jig made from a piece of ⅛-in.-thick hardboard and a 1-in.-wide strip of ½-in.-thick MDF. Take a piece of hardboard that's long enough and wide enough for the task at hand, and use a miter box and handsaw or a powered miter saw to cut it to the desired angle **(A)**. Cut a ⅛-in.-wide, ¼-in.-deep groove centered on the wide face of the MDF strip, and glue the non-angled edge of the hardboard piece into the groove **(B)**. Because the fence overhangs both faces of the hardboard, you can flip the marking gauge over to draw an angle that slopes in either direction.

A dovetail-marking gauge is terrifically useful for laying out dovetail joints to be cut by hand **(C)**. Start by sawing a short fence piece from ½-in.-thick stock and a small wedge-shaped piece from ⅛-in.-thick hardboard. (Save the angled hardboard offcuts.) Each long side of the wedge should slope at the same angle: a taper that's between 6 and 10 degrees.

Draw one centerline to bisect the length of the wedge and another to bisect its width. Clamp the fence atop the wedge, using the angled scrap you saved earlier, to align them **(D)**. Put a couple of drops of thin cyanoacrylate glue on the seam between the parts, and spray the joint with accelerator to set the glue immediately. Then use a couple of small nails or wood screws to strengthen the joint.

[TIP] **To make a more durable angle or dovetail gauge, make it from thin plastic or sheet brass or aluminum.**

A

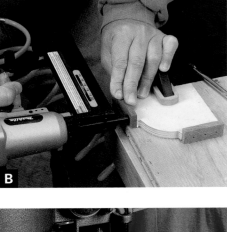

B

C

D

E

Corner and Part Templates

A corner-marking jig is useful for laying out cuts for rounding or shaping the corners of square or rectangular tabletops, or the top edges of a bookcase or other cabinet project **(A)**. First, use a square of ½-in.-thick plywood or MDF to make a template for the corner shape you wish to create, whether it's a quarter-radius curve, an ogee, a cove, or whatever. Saw the shape out with a jigsaw or bandsaw, and clean up the cut with a file or sanding block. Glue and nail a couple of 1-in.-wide strips of solid stock to the edges of the square on either side of the shape **(B)**. The strips serve as stops that will accurately locate the template against the square corners of your project.

Another kind of marking jig provides a template for laying out multiple identical parts, such as curved furniture legs or other complex parts. Make this jig by first drawing the shape of the desired part on ⅛- or ¼-in.-thick hardboard **(C)**. Cut the part out slightly oversized on the bandsaw, scrollsaw, or jigsaw and then clean up the sawn edge by sanding to the line on a disc sander for convex edges and a drum sander for concave edges **(D)**. You can use the template as is, or you may wish to glue and nail a fence strip to one end for parts that must be laid out in a particular orientation relative to the wood's grain **(E)**.

[TIP] Drafting tools, such as circle and arc templates and French curves, will help you draw smooth, flowing lines for good-looking curvaceous furniture and cabinet parts.

Story Sticks

The story-stick method of setting up machines to cut parts is simple, but powerful: All part dimensions—the height of a bulkhead, length of a drawer front, etc.—are marked on a single stick, which is then used to set up fences, stop blocks, etc., on your saws and other machines.

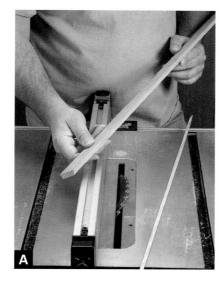

Any long, straight piece of solid stock or thin plywood makes a good story stick. You can improve on the traditional story stick by adding a self-stick measuring tape to it. To size the stick, cut it at least as long as the longest part in your project. Then saw a 45-degree bevel on one edge of the stick on the tablesaw **(A)**. Apply the self-stick tape to this beveled edge, locating the zero mark just a hair in from one end. (If you're right-handed, you'll want to use a tape that reads right to left.) Using a miter gauge on the disc or stationary belt sander, trim the end of the stick until it's perfectly flush with the zero mark on the tape **(B)**. This ensures that the tape will read correctly when butted up against a fence or stop.

Now you're ready to transfer all your project part dimensions to the story stick **(C)**. The idea is to write the name of the part and the relevant dimension (thickness, height, etc.) right on the stick, with a line or arrow pointing to its exact dimension on the tape. Now the stick is ready to use for setting up cuts on machinery **(D)**, laying out the position of hinges and hardware, or aligning parts during assembly.

[TIP] **You can use short story sticks to record all the part dimensions, joinery layout, and hardware positions of projects built without machine tools.**

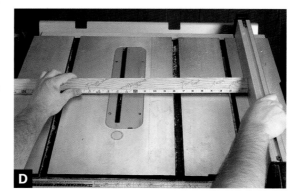

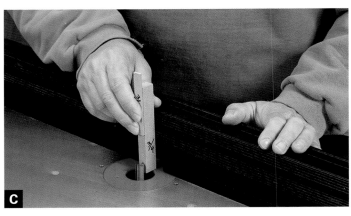

Depth Gauges

The majority of tablesaws, shapers, and other shop machines lack scales to help set their depth of cut accurately. Individual shopmade gauges, made from dimensionally stable hardboard or MDF, help you quickly and accurately set the height of a blade or cutter, as well as the distance between a fence and cutter.

A set of gauges can be made easily on the tablesaw. Rip a series of ½-in.-wide, 5-in.-long gauge strips to various common thicknesses such as ⅛ in., ¼ in., ⅜ in., ½ in., ¾ in., etc. Make sure each strip is *exactly* the desired thickness by measuring it with a caliper. Also cut a ½-in. x 3½-in. x ⅛-in. hardboard stop strip for each one. Glue a stop strip atop each gauge strip, with about ¾ in. of the stop hanging over the end. The stop strip can be "clamped" in place using tape until the glue dries (A). Use the stop strip end of the gauge to set the blade or cutter height (B), and the other end to set the distance between the cutter and the fence or stop (C).

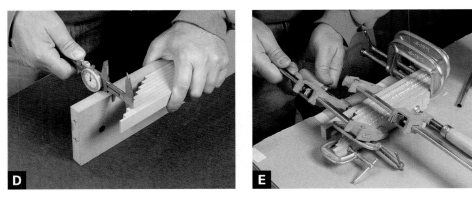

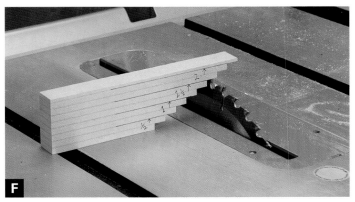

You can also make a handy cutting-depth gauge by gluing up a stack of strips cut from ¼ in.- and/or ⅛ in.-thick hardboard. On the tablesaw, cut nine strips 8 in. long and 1 in. wide. Now hold or clamp the strips together (D), to make sure the hardboard gauge will show accurate measurements. If necessary, use paper shims between the layers of hardboard to adjust the distances. Apply glue and clamp the strips together, extending each strip about ⅜ in. beyond the one below it (E). After the glue cures, scrape and sand the sides of the gauge smooth, cut one end square, and note the height settings with a marker. To use the gauge, hold it flat on the machine table and slowly raise the blade or bit until it touches the desired strip (F).

Setup Gauges

Most setup jigs and gauges take some effort to make. But did you realize that you can automatically create a set of time-saving setup gauges just by saving the wood pieces you use to take trial cuts? This is especially useful for router and shaper operations using bits and cutters that form complex cope-and-stick or miter-lock joints.

The first time you set up a cut (or series of cuts), use the information provided by the cutter manufacturer to set cutting height and fence distance with a ruler or caliper **(A)**. Using a pile of scrap stock machined to the same dimensions as your workpieces, take a series of test cuts, tweaking machine settings until you get a perfect cut **(B)**. In the case of joinery bits or shaper cutters, check the fit of test pieces to ensure a tight, properly aligned joint **(C)**. Label the test piece(s) with all the relevant cut information, including which bits and bearings were used, and in what orientation. The next time you need to make the same cut, use the trial piece(s) as setup gauges to quickly position bit height, fence distance, etc., without having to take a whole series of trial cuts again **(D)**.

▶ See *"Recording Jig Instructions"* on p. 43.

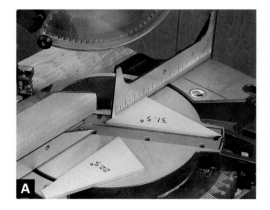

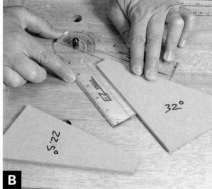

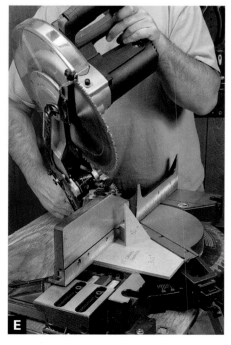

Angle Gauges

Ever get tired of taking trial cuts each time you need to set up an odd-angle cut on the sliding-compound miter saw or radial-arm saw? It only takes a few minutes to make an angle-setting gauge that'll help you reset your saw's miter and/or bevel settings the next time you need to repeat that cut.

Make the gauge from ⅛-in. or ¼-in.-thick hardboard or tempered Masonite by simply cutting out a triangle with one side sloping at the desired angle **(A)**. For best accuracy, use a protractor or bevel gauge with a scale to make sure your cut is true **(B)**. Write the angle degrees on the gauge and mark which corner of the triangle the angle describes. You can also use this kind of gauge to set angle cuts on a tablesaw. To make a gauge easier to use for bevel settings, cut it out of ¾-in.-thick MDF and then drill two shallow holes along one edge and glue a pair of ½-in.-dia. disc type magnets into them. The magnets hold the gauge firmly on the sawblade (or machine table) as you tilt the blade **(C)**.

To make a gauge for a compound-angle cut, combine two triangles—one that you've cut to the desired miter angle, and a smaller one cut to the desired bevel angle. With the larger triangle lying flat on the workbench, glue the smaller triangle on top, supported by a small glue block. Orient the smaller triangle square to the large triangle's angled edge **(D)**. This gauge is very handy for accurately setting up complex crown molding cuts **(E)**.

[TIP] Record all information about crown molding cuts (type of molding, inside or outside cut, etc.) on the angle-setting jig.

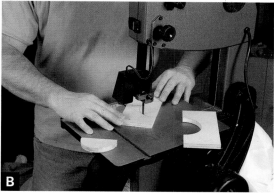

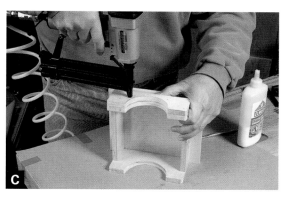

Router Cradle

You can make the onerous task of changing router bits much easier by building a cradle that will support your router and lock its arbor during the process. The basic router cradle, shown in the drawing below, works with any router that has a removable base.

Build the cradle from ½-in.-thick plywood, sizing it to suit your router. Start by measuring the diameter of your router's body at the top and bottom ends. Cut out a strip long enough for the cradle's two rectangular end pieces, and use a compass to mark out half-circles that are slightly larger than your router's top and bottom diameters **(A)**. Also cut and mark the spacer piece for the bottom end of the cradle. Saw the half-circles out with a jigsaw or bandsaw **(B)**.

Next, cut out the cradle's baseplate and sides on the tablesaw. Make the sides 2 in. to 2½ in. wide and long enough so that the circular cutouts in the end pieces support the router motor at its top and bottom ends. Make the base at least 2 in. wider than the length of the end pieces to give the cradle good stability. Glue and nail the cradle parts together, adding the spacer to the end piece as shown **(C)**.

(Text continued on p. 62.)

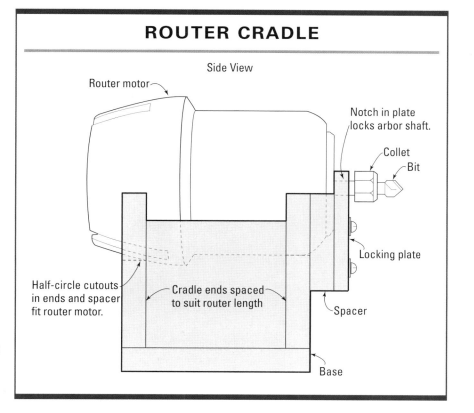

ROUTER CRADLE

Side View

Router motor

Notch in plate locks arbor shaft.

Collet

Bit

Locking plate

Half-circle cutouts in ends and spacer fit router motor.

Cradle ends spaced to suit router length

Spacer

Base

One neat feature of the cradle is its built-in arbor lock, which allows you to loosen or tighten the router's collet using only a single wrench. Make the lock plate from a piece of ¼-in.-thick phenolic plastic or aluminum the same width and length as the cradle's spacer. Then cut a square notch, centering it in the edge of the plate, and sizing it to match the innermost arbor nut (D). Initially cut the notch undersized and then carefully enlarge it until it fits the arbor nut snugly. Round the top corners of the notch slightly for easier insertion of the arbor nut. Mount the plate to the spacer with four washer-head screws inserted through slightly oversized holes, which will allow adjustability (E). Don't tighten the screws fully until you're sure that the cradled router engages the lock properly.

To use the jig, simply remove the router motor from its base and set it into the cradle, guiding the arbor nut into the locking plate notch (F). Try not to overtighten the collet when installing bits. Excessive force is unnecessary and tends to damage the locking plate, making bit changes more difficult.

[TIP] If your router has a built-in arbor lock, add a plywood finger or dowel rod to your router cradle to depress the locking lever or button during bit changes.

Fences and Guides

WOODWORKERS USE FENCES practically every time they work in the shop; a tablesaw, jointer, or crosscut saw would be downright dangerous to use without one. Depending upon the machine and operation, a fence or a guide may serve to steer a board past a blade or cutter, or to position a workpiece for a perfectly square or angled cut. It may also guide a piece for accurate drilling of holes with a drill press. Although most fences and guides are designed to work with straight stock, there are also curved-fence jigs intended specifically to handle (you guessed it!) round and curved work. Fences for guiding portable power tools are also very useful, as discussed in Section Eight.

Although most woodshop machines come equipped with a fence of some kind, there are many times when it's best to replace or augment the tool's standard fence. This includes specific operations such as guiding or supporting workpieces that are longer or larger than usual, or adapting a machine for cuts it's not normally intended to make. For example, standard ripping operations require that boards travel parallel to the tablesaw blade. However, a cove-cutting fence guides the work over the blade at an angle, which allows you to create a wide range of profiles using a standard blade.

Parallel Fences

➤ Utility Rip Fence (p. 65)
➤ Cove-Cutting Fence (p. 66)
➤ Mortising Fence (p. 69)

Auxiliary Fences

➤ Box Fence (p. 70)
➤ Long Rip Fence (p. 72)
➤ Router Table Zero-Clearance Fence (p. 73)

Short Fences

➤ Short Rip Fence (p. 74)
➤ Resaw Fence (p. 75)
➤ Router Table Split Fence (p. 76)

Angled Fences

➤ Frame Miter Jig (p. 77)
➤ Adjustable Miter Jig (p. 78)
➤ Crown Molding Fence (p. 79)
➤ Pivoting Fence for Drill-Press Table (p. 81)

Curved Fences

➤ Disc Fence (p. 83)
➤ Vertical Disc Fence (p. 84)
➤ Arc-Cutting Fence (p. 85)
➤ Curved-Part Shaping Fences (p. 86)

Dowel Guides

➤ Dowel-Splitting Guide (p. 88)
➤ Dowel-Pointing Guide (p. 89)
➤ Dowel-Sizing Jig (p. 90)

A box fence fastens easily over your tablesaw's stock rip fence and lets you bury a dado blade in the fence face for cutting rabbets.

Cutting perfect miters for 90-degree corners is a snap with a picture frame jig mounted to your crosscut saw.

When a machine's stock fence isn't quite up to the task at hand, an auxiliary fence can add length or features to make a job easier and safer. For example, a box fence (see the top photo) not only lets you safely bury part of a cutter when you're making rabbet joints, but it features a T-track for slide-on guards, hold-downs, and other accessories.

Another kind of add-on, partial fences, are short but big on function. On a router table or shaper, a split fence is an essential accessory, allowing you to set the distance between fence faces to suit the diameter of the bit or cutter. When you're resawing on a bandsaw, a short fence helps you get truer cuts, and when you're ripping squirrelly stock on a tablesaw, it can prevent dangerous kickbacks.

Angled and pivoting fence jigs are designed to handle another set of useful woodshop chores, including setups for quick and perfect picture-frame miters (see the bottom photo at left), sawing miters at a wide range of angles, or mastering the fine (and complicated) art of cutting crown molding on a crosscut saw. The pivoting drill-press fence in this section is a terrific drilling aid, and probably one of the most useful jigs in this entire book.

Although you may not use one every day, a curved fence is a lifesaver when specific tasks require cutting or shaping round and curved parts or panels. The novel arc-cutting fence I'll discuss is a particularly handy jig for cutting arched upper door rails, as shown in the photo below.

Finally, this section includes a number of very useful fence jigs specifically made for cutting and shaping round wood dowels. Included are specific jigs for splitting dowels in half on a bandsaw and for shaping chamfers and points on the ends of dowels using a disc sander. There is also a very practical jig for reducing the diameter of slightly oversized dowel stock.

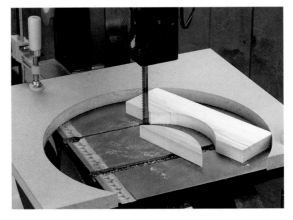

You can quickly cut an entire kitchen's worth of arched door rails using this simple circular bandsaw fence.

Utility Rip Fence

Whether you're looking to replace that rickety old tablesaw fence or whether you just need a fence with a wooden face that you can bury a blade in, this utility rip fence works well for standard rip cuts as well as cove cuts. To build it, start by cutting a 2½-in.-wide fence bar from straight-grained 1-in.-thick solid wood stock. Size the bar to be at least 8 in. longer than the width of your saw table (plus fence rails, if present). Make it longer if you routinely cut big panels. Using ½-in.-thick plywood, make a 30/60/90-degree triangle, the hypotenuse of which is slightly longer than one-third the length of the fence bar. Glue this triangle to the right-hand side of the fence, reinforcing the joint with a glue block strip about ¾-in. square.

The fence bar mounts to a 1¾-in.-wide by 16-in.-long crossbar cut from 2x4 lumber. Bore holes through the fence and crossbar, centered on the crossbar and about 4 in. from the fence's front end. Then connect the parts with a 5-in.-long carriage bolt **(A)** secured by a hand knob on the underside.

With the saw unplugged, locate the fence bar within easy measuring distance from the blade, then clamp the crossbar to the front edge of your saw table (or, if practical, to its existing fence rail) **(B)**. Raise the sawblade and carefully adjust the angle of the fence bar until it's precisely parallel with the blade, measuring the distance at both front and back edges of the blade **(C)**. Drill a hole for a wood screw near the far corner of the triangle and then drive it into the crossbar to secure the fence's position parallel to the blade **(D)**. For cove-cutting work (as described in the next photo-essay), remove the screw and readjust the fence angle as desired.

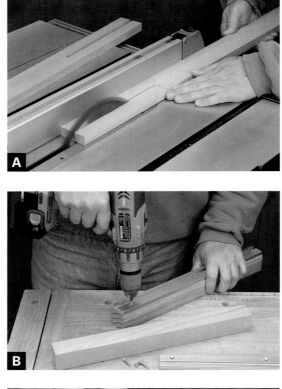

Cove-Cutting Fence

Cutting a cove profile into molding or a cabinet part is easy if you have a large shaper and big, expensive cutters. But you can also cut a wide range of cove sizes and shapes on the tablesaw with a regular sawblade and an angled fence. The cove is created by passing the stock over the blade at a skewed angle; the greater the angle, the larger the radius of the cove, as shown in the drawing on the facing page. However, unlike shaper cutters, which can hog out a profile to full depth at once, tablesaw coves are cut a bit at a time by raising the sawblade about ⅛ in. between passes.

While you can get away with cove cutting using only a straight board clamped to the saw table as a fence, the cove-cutting fence created here has twin rails that securely guide the workpiece over the blade, making the process both easier and safer. Using ¾-in.-thick straight-grained lumber, cut a pair of fence rails 2¼ in. wide and at least 36 in. long. In one end of each rail, rout or saw a centered ¼-in.-wide slot, 10 in. long **(A)**. At the other end of each rail, bore a centered ¼-in.-dia. hole, about 1 in. from the end. The rails are connected by a pair of crossbars cut from 2x4 lumber, each 1½ in. x 1½ in. x 16 in. A 16-in.-length of T-track is screwed to the top of each crossbar **(B)**, and then four 1½-in.-long, ¼-in.hex-head bolts and hand knobs are used to attach the rails to the bars **(C)**.

Before using the cove fence, unplug your tablesaw and lower its blade fully. Clamp the crossbar at the non-slotted end of the jig's rails to your rip fence rail or to the front edge of the saw table, positioning the bar to the left of the blade's line of cut **(D)**.

(Text continued on p. 68.)

COVE CUTTING ON THE TABLESAW

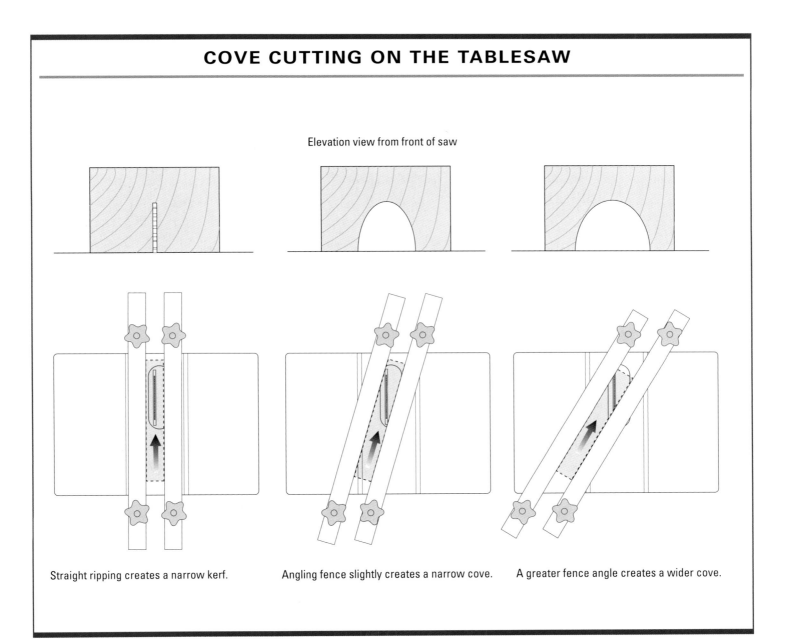

Elevation view from front of saw

Straight ripping creates a narrow kerf.

Angling fence slightly creates a narrow cove.

A greater fence angle creates a wider cove.

Raise the sawblade to a height that matches the depth of the desired cove, and set the position and angle of the right-hand rail so that both the width of the cove and the distance between it and the edge of the workpiece are as desired **(E)**. Clamp the rear cross member to the rear edge of your saw table and secure the right rail by tightening the hand knobs at the front and rear. Place your workpiece against the right-hand rail, and position the left-hand rail against the opposite edge of the work. Then lock the rail on the crossbars **(F)**.

Lower the sawblade so that it's only about ⅛ in. above the table. Now you're ready to start cutting a cove. Using push blocks and sticks, slowly pass the workpiece over the blade. Then raise the blade no more than ⅛ in., and take another pass **(G)**. Repeat the process until your cove is almost complete. Then raise the blade a scant ¹⁄₃₂ in. to ¹⁄₁₆ in. for the final pass. This will leave a relatively smooth surface that will clean up with minimal sanding. If you want to repeat the particular cove profile in the future, cut a setup triangle that matches the angle of the fence rails, and draw the cove shape on it for reference **(H)**.

Mortising Fence

A square-chisel mortising attachment on a drill press provides a faster way of cutting mortises than chopping them out with a mallet and chisel. This sturdy fence features a handy sliding wedge design that sets the fence relative to the mortise chisel, while keeping it parallel. The sliding-wedge mechanism is made up of two mating 18-in.-long tapered pieces cut from ½-in.-thick plywood using a taper jig on the tablesaw. (See the bottom photo on p. 109.)

The wider strip tapers from 2¾ in. down to 1¾ in.; the narrow one, from 1½ in. to ½ in. On the wide taper strip, rout a ¼-in.-wide, 12-in.-long slot centered along the tapered edge **(A)**. Glue and nail the narrow taper (tapered edge facing in) centered at the back edge of a 24-in.-long, 7½ in.-wide base cut from ½-in. or ¾-in.-thick plywood.

Make a 24-in.-long fence from two 3½-in.-wide pieces of straight-grained stock glued up square to one another. Cut slots for plate joinery biscuits in the back edge of the fence and in the non-slotted edge of the wide taper, and glue the assembly together **(B)**. Glue an 18-in. x 3¾-in. spacer cut from ½ in. ply to the back of the fence. Attach a length of T-track to the top of the fence **(C)**, and use it to mount the hold-down assembly that keeps the stock from lifting during mortising. Now butt the wide taper against the narrow one on the base, and install a pair of ¼-in. hangar bolts (set 6 in. apart) through the slot into the base. A pair of hand knobs on these bolts locks the fence assembly in place. With the base clamped down to the drill press table **(D)**, loosen the hand knobs and slide the wide taper side to side to move the fence face in or out relative to the mortising chisel.

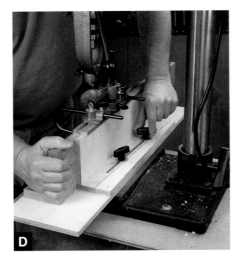

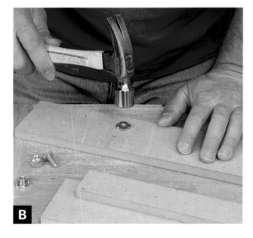

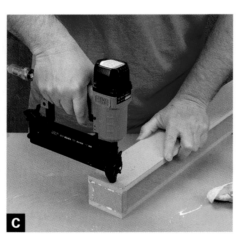

Box Fence

A box-style auxiliary fence fits over your table-saw's existing rip fence and attaches quickly whenever you need to bury part of the sawblade or dado head to make a cut such as a rabbet. Constructed from ½-in.-thick MDF or plywood, this useful fence is built like a long, U-shaped channel with an open bottom. Size the box so that its inside dimensions are just ½₂ in. or so greater than the height and thickness of your rip fence **(A)**, so it'll fit over it snugly. Cut the sides long enough to extend past both ends of your fence and make the top a little shorter, to clear the fence's locking lever when it's in the "up" (unlocked) position. On the side of the box that faces away from the blade, drill a couple of holes about 1 in. down from the top edge, and drive a couple of T-nuts into the holes **(B)** (on the inside facing side). The nuts are for a pair of thumb-screws that will clamp the box firmly to the rip fence. Now glue and nail the sides and top together, fitting an end cap to the far end of the box to reinforce the structure **(C)**. To make it easier to attach hold-downs or guards to the box fence, screw a length of T-track to the top **(D)**.

The box fence should slip easily over your rip fence, requiring just a quick tightening of the two thumbscrews to keep it firmly in place **(E)**. To keep thin stock from sneaking underneath, slide the box fence all the way down until it completely contacts the saw table.

[VARIATION] You can make a box-style auxiliary fence even more versatile by screwing on accessory strips as they are needed. For example, installing a small strip of hardboard into a groove cut into the face of the box (A) keeps thin stock, such as plastic laminates, veneers, and thin plywoods, from lifting up off the saw table as you cut it. Locate the groove high enough on the fence so that the sawblade will clear the hardboard strip.

Another time an add-on strip may be useful is when you need to bevel an edge at an angle greater than the maximum blade tilt of your saw, which is typically 45 degrees. In this case, screw a strip to your box fence's face to elevate one edge of the work (B). The surface of the strip is beveled at an angle that augments the tilt of the sawblade so it can cut the proper angle. For example, to rip a 58-degree bevel on the work, the sawblade is tilted to 45 degrees and the strip is beveled at a 13-degree angle (45 + 13 = 58).

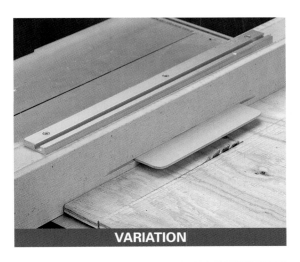

VARIATION

VARIATION

A

B

C

Long Rip Fence

Outfeed tables and support rollers are fine for most ripping jobs. But long, narrow wood strips and thin sheet stock often need more support to glide smoothly without hanging up on outfeed supports. This long ripping fence not only improves stability when you're sawing large panels, but it also provides a small support surface ahead and behind the saw table to support thin and narrow workpieces **(A)**. To make the fence face, start by ripping a strip of straight-grained stock two feet longer than your saw table and the same height as your rip fence **(B)**. Cut a couple of supports from ½-in.-thick plywood, making them 12 in. long and 4 in. to 8 in. wide. Glue and screw the supports to the underside of the fence face at the ends **(C)**, making sure that the supports will clear your rip fence rails and cursor assembly. If necessary, cut a notch in one or both supports to provide clearance. It's a good idea to chamfer the leading top edge of the outfeed support to prevent stock from hitting it.

You can attach the long fence to your rip fence using a T-slot, if your fence has one, or you can simply clamp or screw it **(D)** in place temporarily. Before ripping, make sure that the support tables are level with the saw table.

D

Router Table Zero-Clearance Fence

The standard fence on your router table is fine for most operations. But when you're working with really big bits or short workpieces, or routing wood with curly figure or other problematic grain, a stock fence may not do the trick. By eliminating any open space around the bit, a zero-clearance fence prevents chatter and splintering while supporting the workpiece fully as it's pushed through the cut. To make this fence, cut a strip of ½-in.-thick MDF, plywood, or solid stock that's as high and as long as your router table's regular fence. At the center point of this strip, mark the profile of the router bit you intend to use **(A)**. With a jigsaw or coping saw, cut out the bulk of the area inside the profile, staying about ⅛ in. inside the line **(B)**. Completely cut out the area around the pilot bearing, if your bit has one.

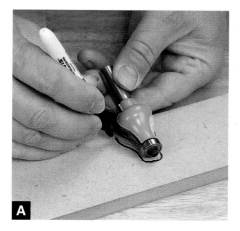

After screwing or bolting the zero-clearance fence to your regular fence, clamp a piece of thick scrap to the fence face at the bit location. If necessary, clamp one end of the fence to the table to prevent side-to-side movement. Carefully switch the router on, and swing or slide the fence back **(C)** until the bit has penetrated the zero-clearance fence to the extent that it will when you rout with it. Then, with the router off, lock the fence down and remove the scrap block. Turn the router back on and make sure the bit isn't rubbing excessively on the fence. If it is, you may need to back the fence off slightly, or repeat the through routing with the block. Even though you need to make a separate zero-clearance fence for each bit and setup, the clean and safe cuts you get **(D)** make it worth the trouble.

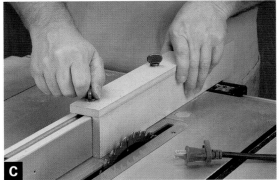

Short Rip Fence

European tablesaws usually feature a rip fence that allows fore-and-aft adjustment. One use for this is to provide clearance just past the blade, to minimize the risk of kickback. You can get the same effect by adding a short auxiliary fence to the face of your regular rip fence. Cut the short fence from $\frac{1}{2}$-in.-thick MDF or plywood that's at least as high as your regular rip fence and half to two-thirds the length. If your rip fence features a T-slot along the top, then make the face $\frac{1}{2}$ in. wider than the height of your fence, and cut a mounting strip that's as wide as your rip fence. Bore a couple of holes in this strip, centering them on the T-slot, and then glue and nail the strip along the top edge of the fence face (A). To reduce friction, add a layer of self-stick UHMW (polyethylene) tape to the face.

> See *"Slippery Plastics"* on p. 20.

Press the tape down firmly with a roller or a block of wood (B).

Mount the short fence to your stock fence by installing a couple of T-nuts and thumbscrews through the holes you bored. Slip the nuts into your fence's T-slot, and slide the fence forward until its rear edge is forward of the back teeth of your tablesaw blade (C). If you rip a piece of "squirrely-grained" wood, the short fence allows clearance to prevent the ripping from pinching and binding on the blade, thus preventing burning and kickback (D).

Resaw Fence

A popular method of resawing thick stock on the bandsaw is to use a single-point fence that allows you to adjust the feed angle of the stock as you saw. This way you can compensate for "blade drift," which can cause erratic cuts. The face of this resaw fence is actually a ½-in.-dia. dowel that was cut in half **(A)** using a dowel-splitting jig.

▶ See "*Dowel-Splitting Guide*" on p. 88.

Cut out the fence's triangular support pieces from ½-in.-thick plywood. Size the largest triangle to suit your bandsaw's maximum height capacity; the two smaller triangles provide side support. Glue the half-dowel to the edge of the large triangle using masking tape as a clamp. Cut a 3-in. to 6-in.-wide jig base from ½-in.-thick plywood or MDF, making it long enough to span two-thirds of your bandsaw table. Fasten a cross member to the end of the base that butts up against your saw's table or rip fence rails. Glue the triangular fence to the base **(B)**, so that the half-dowel will align with the saw-teeth. (It's best to resaw with a ½-in. to 1-in.-wide blade.) Install screws or nails through the base to secure the fence.

Clamp the jig's cross member to your saw **(C)**, setting the distance between the dowel and blade to the desired resaw thickness. Mark an easily visible pencil line on the edge of your stock that matches the resaw dimension; you'll use this line to guide the stock as you saw. For best results, use a wheeled hold-down to press the stock securely against the fence as you work **(D)**.

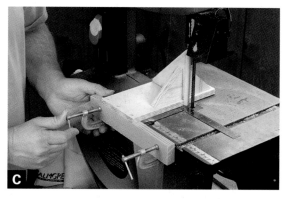

▶ See "*Wheeled Hold-Down*" on p. 241.

[TIP] Before resawing a thick board, make a V-shaped mark completely across its edge to help you realign the grain of adjacent leaves later, if desired.

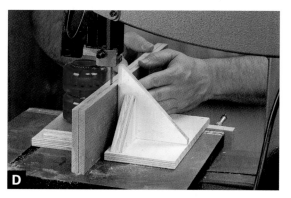

Router Table Split Fence

Wooden split-fence faces that attach to your regular router-table fence let you adjust the fence opening to support workpieces as closely as possible to a bit. Cut two fence faces from straight-grained stock, ripping each as wide as the height of your router table fence and half its length. Back-bevel the inside-facing ends of each piece, leaving a slightly blunt end rather than a sharp tip. To allow adjustment, a T-slot is routed into the back side of each face, about 1½ in. from the bottom edge. To make the T-slot, first rout or saw a ¼-in.-wide by ⁵⁄₁₆-in.-deep groove. Then finish routing the groove using a T-slot bit in a router fitted with an edge guide **(A)**.

For mounting guards and hold-downs, rout another T-slot in the front of each face, ¾ in. to 1 in. down from its top edge **(B)**. Bore four ⁹⁄₃₂-in.-dia. mounting holes in your router-table fence, positioning them 2 in. from the ends and 2 in. from the center of the fence. Insert a short T-bolt though each hole, slide on the wood faces, and then secure them with hand knobs. For the best workpiece support and the cleanest cut, set the beveled ends as closely as possible to the bit **(C)**.

You can also use a split router-table fence to square and straighten the edges of stock, much as a jointer does. Add a pair of identical shims behind the outfeed fence, placing one at each end **(D)**. The thickness of the shims should match your desired depth of cut—typically between ¹⁄₃₂ in. and ¹⁄₁₆ in. Adjust the shimmed fence flush to the cutting edge of a straight bit chucked in the table router, and you're ready for jointing.

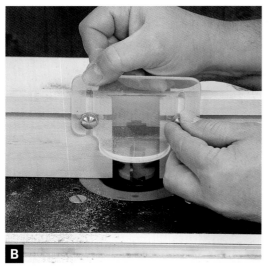

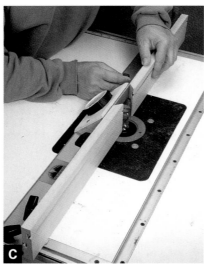

Frame Miter Jig

Unless your power miter saw is really accurate and easy to set, it's more efficient to cut miters for frames, shadow boxes, and such by leaving the saw set for a square cut and using this frame miter jig **(A)**. To make the jig, first cut a base-plate from ½-in.-thick plywood or MDF, making it as long as your saw table and a couple of inches wider than the saw's crosscut capacity. Next, rip two 1½-in.-wide wood fence strips from straight-grained stock and miter the end of each at 45 degrees **(B)**.

Position and clamp the jig base atop your saw. Then, with the saw set at 90 degrees, cut a kerf part way through the base. Use a drafting triangle to align one of the fence strips at 45 degrees to the kerf, with its tip just touching the kerf, and then glue and nail the strip in place. Using a large, accurate framing square, position the other fence strip at 90 degrees to the first **(C)**, and glue and nail it down. To strengthen the jig's base, glue a triangular block of wood to the base at the end of the kerf. Finally, glue and nail a couple of cleats to the back edge of the base to allow the jig to be clamped to the saw's fence. Whenever you're clamping the jig to the saw, set the saw-blade in the kerf to position the jig **(D)**.

[TIP] For perfect picture frame joints, parallel frame members must be exactly the same length as well as being accurately mitered to 45 degrees.

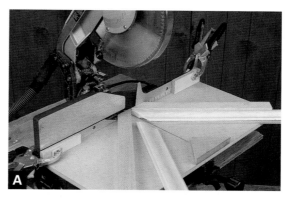

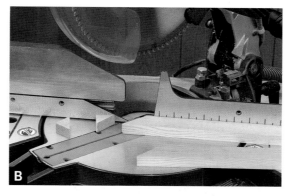

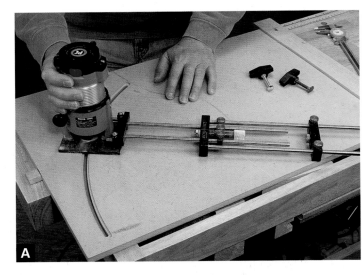

Adjustable Miter Jig

This pivoting-arm fence jig eliminates the need for time-consuming saw adjustments in setting up odd-angle miter cuts on a crosscut saw. Make the base from ½-in.-thick MDF cut to 25 in. long and 2 to 3 in. wider than your saw's crosscut capacity. Mark a cutline across the baseplate 5 in. from the right-hand end. Drill a hole for a fence pivot bolt 2 in. to the left of the line and 1½ in. from the back edge of the base. Counter-bore the underside of the hole to recess the head of a carriage bolt. Next, rout the curved slot through the base, using a router outfitted with a ¼-in.-dia. straight bit and a circle-cutting guide. Set the guide for a 16-in. cutting radius, and swing it from the arm's pivot hole. Widen the slot on the underside of the base using a ½-in.-dia. straight bit set for a ⅛-in.-deep cut **(A)**.

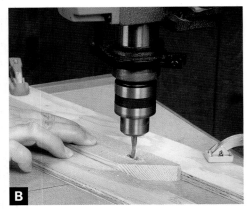

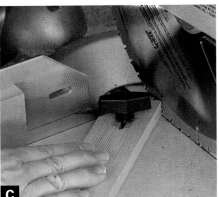

Saw a ¾-in. x 2-in. x 21-in. pivot arm from straight-grained stock, and then miter one end at 45 degrees **(B)**. Then rout two ¼-in.-wide slots about 1 in. long along the axis of the arm, spacing them 16 in. apart. (These allow the pivot arm to move in or out slightly so its tip can remain flush with the cutline.) You can cut the slots with a router or drill press fitted with a ¼-in.-dia. spiral end mill or router bit **(C)**. Secure the pivot end of the arm with a ¼-in. T-bolt and hand knob. Glue and nail a cleat to the back edge of the base for clamping the jig to the saw's stock fence. Cut a kerf through the base, and use it to align the jig to the saw when you're mounting it **(D)**.

[TIP] PSA sandpaper stuck to a fence face helps prevent workpieces from slipping as they're cut.

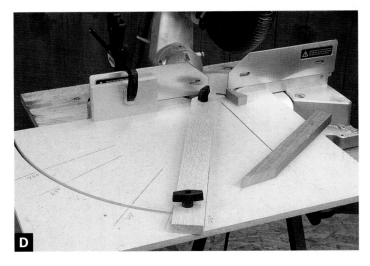

Crown Molding Fence

One of the most challenging woodworking jobs is cutting and installing crown molding. Designed to be mounted where a wall or the tops of cabinets or soffits meet a ceiling, this wide molding is installed at an angle. It's this angle that makes crown molding devilish to cut—and to figure out how to orient the pieces when making left- or right-hand cuts on inside or outside corners can be maddening. You can lay the molding flat on a powered compound-miter saw to make the cuts, but the saw's miter and bevel settings must be changed for each cut. It's much easier to build and use a fence jig that orients and supports the molding during cutting.

Because crown molding comes in many sizes and styles, you have to measure the particular molding you're using. Lean a piece of molding upside down against a square block **(A)**, and check both its width and height, as these dimensions usually aren't the same. Note that the correct "right-side-up" orientation for installed crown molding is with the broad ogee section closer to the ceiling **(B)**.

The basic jig is a plywood or MDF channel consisting of a base strip and two side strips, each cut 20 in. to 24 in. long (longer if you'll be cutting really long pieces). Make the inside width and height of the channel so that it matches the dimensions of the molding you measured. Glue and nail the channel together **(C)**, but leave a few inches free of nails near the ends. Then saw an opposing 45-degree miter on each end of the channel **(D)**.

In use, the jig is placed against the miter saw's left-hand fence, with the end flush with the line of cut. The molding is then set upside down inside the channel **(E)**. Depending on the particular cut you're making (inside or outside, right-hand or left-hand corner) there are three factors to consider: First, the saw's miter angle is set at 45 degrees to either the right or to the left. Secondly, either the long side or the short side of the jig is placed against the saw fence. Last, the molding is set into the jig

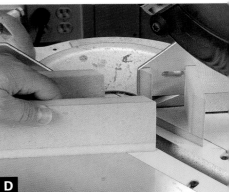

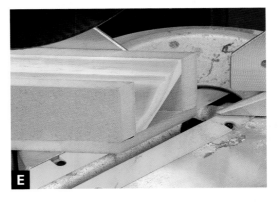

(Text continued on p. 80.)

CROWN MOLDING FENCE

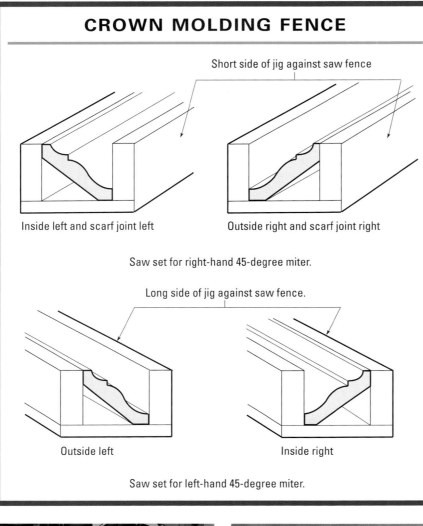

Short side of jig against saw fence

Inside left and scarf joint left

Outside right and scarf joint right

Saw set for right-hand 45-degree miter.

Long side of jig against saw fence.

Outside left

Inside right

Saw set for left-hand 45-degree miter.

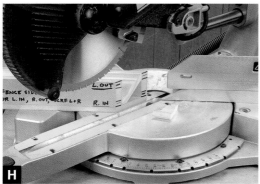

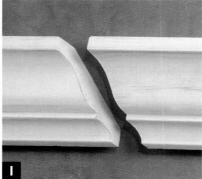

sloping either towards or away from the saw fence. The drawing at left shows the orientation of the molding, saw and jig for each of the four basic crown molding cuts, as well as for scarf joints, discussed below. (For my descriptions here, "right-hand" and "left-hand" refer to the relative ends of mating workpieces as viewed when installed. For example, a "right-hand outside corner" actually refers to the cut made on the left-hand end of the right-hand workpiece.)

To cut an outside right-hand corner, set the saw to cut 45 degrees to the right, place the short side of the jig against the fence, and set the molding to slope away from the fence (F). Press the molding firmly down into the channel, holding the channel against the saw's fence as you cut (G). To make either an inside-right or an outside-left corner cut, flip the jig around so that its long side is against the fence, and readjust the miter setting for 45 degrees to the left (H). To help you remember how to use the jig for all basic cuts, write the information directly on the jig after you perform each cut for the first time.

You can also use this crown molding fence jig to cut scarf joints—sometimes necessary when joining two lengths of molding end-to-end (I). In a scarf joint, the ends of the adjacent pieces overlap each other at 45 degrees to prevent the kind of gap you might see if the square-cut ends were simply butted together.

[TIP] **When installing crown molding to out-of-square walls or cabinets, adjust the saw's miter setting slightly to compensate.**

Pivoting Fence
for Drill-Press Table

This pivoting-arm fence jig makes most boring operations on the drill press quicker and more precise. The jig's adjustable fence and end stop let you locate holes relative to the width and length of the workpiece with repeatable accuracy **(A)**. Start by cutting the jig's 16-in. x 24-in. base from a piece of ¾-in.-thick plywood. Mark a point in the exact center of the base, which will be used to align the jig with the center of the drill press chuck later.

Locate a pivot point for the jig's fence, ¾ in. from the left-hand edge of the base and 2 in. from the rear edge. With a beam compass, draw a 23-in.-radius curve on the right edge of the base, extending from the back edge to 3 in. short of the front edge. Then drill a hole at the pivot point. Saw out the curved section and the notch with a jigsaw **(B)**, and then smooth the edge with sandpaper. To make the jig mount quickly and easily to your drill press table, glue and nail a couple of positioning blocks to the bottom of the base **(C)**, making sure the blocks accurately align the center point of the base with the center of the drill chuck.

Make a 2-in. x 28-in. wood fence arm from straight-grained 1-in.-thick stock. Bore two holes through its thickest dimension, spacing them 23½ in. apart. These will accept the two carriage bolts, one of which serves as a pivot point, and the other of which locks the arm to the base. For a cleaner look, counter-bore the holes for the carriage-bolt heads.

To provide clearance for the drill press chuck when you're boring close to the pivot fence, bore a half-hole into the top edge of the fence. Clamp a scrap block to the face of the fence, and bore a 1½-in.-dia. hole about 1 in. deep **(D)**.

(Text continued on p. 82.)

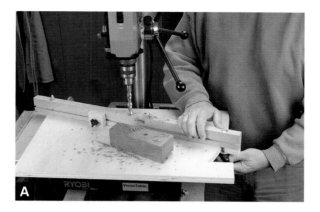

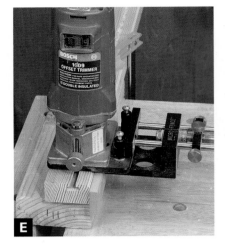

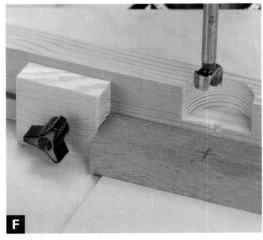

Rout a T-slot into the lower face of the fence to accept the jig's sliding end stop. To make the T-slot, first cut a ¼-in.-wide, ⅜-in.-deep groove, centering it ¾ in. up from the bottom of the fence. To complete the slot, re-rout the groove using a special T-slot router bit **(E)**. Cut a 1¾-in. x 2-in. stop block from ¾-in. stock, and drill it for the T-bolt and hand knob that lock it into the slot **(F)**.

Before attaching the pivot arm to the base, glue a small block cut from ¾-in. plywood to the underside of the arm to act as a spacer for the locking bolt **(G)**. Fit a large fender washer under the hand knob for securing the pivot arm. Use a locking nut to secure the pivot bolt **(H)**.

To make the jig faster to adjust to commonly set distances—say to space holes ½ in., 1 in., 2 in., etc. from the edge of the work—mark the position of the fence at each distance: Measure from the center point of the bit to the fence, and then strike a pencil mark on the base to mark the fence position **(I)**.

[TIP] You can make the fence on a pivot jig slant either to the left or to the right, depending on your personal preference.

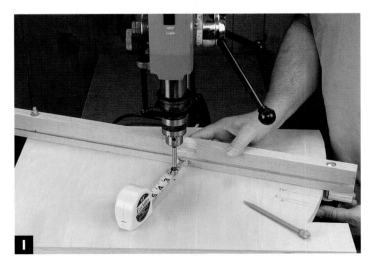

Disc Fence

Functional or decorative routing and drilling on a wooden wheel or disc is easy with this simple fence. Made from an 11-in.-sq. panel of ¾-in.-thick MDF, the jig's V-shaped cutout guides cuts on discs ranging from 2 in. to 12 in. in diameter.

Using a power miter saw, make two 45-degree cuts in opposing directions to create the V, stopping each cut about halfway across the panel **(A)**. To make the fence easier to align during use, use a square to draw a pencil line from the intersection of the cuts to the rear edge of the panel. Then, using the drill press, bore a 1-in.-dia. hole where the cuts meet **(B)**. The hole will provide clearance for a bit when you're routing small discs.

The disc fence is useful on the drill press for center-drilling pre-cut wheels or for boring holes around the perimeter of a disc, say, to make an index plate for a lathe.

▶ See *"Lathe Indexing Plate"* on p. 204.

After positioning the V to align the work as desired, simply clamp the fence to the drill-press table. For evenly spaced holes around the disc, make a mark on the drill-press table and align it with index marks drawn around the edge of the disc **(C)**. The disc fence is also useful on the router table for routing grooves for inlay decorations or for creating decorative shapes **(D)**. To make the fence easier to clamp to a large router tabletop, you can screw it to a piece of 1x2 or 2x4 stock that spans the tabletop.

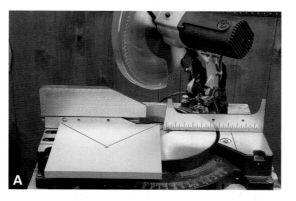

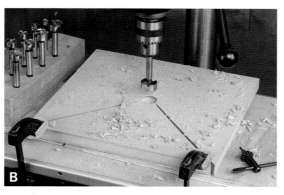

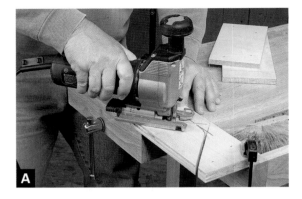

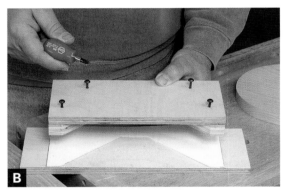

Vertical Disc Fence

This fence jig supports and guides a disc vertically for edge grooving or slicing on a tablesaw or shaping on a router table. To make a jig that'll work with ¾-in.-thick discs between 4 in. and 8 in. in diameter, cut a 4½-in. x 10-in. fence piece from ¾-in.-thick plywood, and then saw out with a bandsaw or a jigsaw **(A)**. From ½-in.-thick plywood, cut two more 4½-in.-wide pieces, a 13-in.-long clamping base, and a 10-in.-long fence cover. The three pieces are sandwiched together with the fence piece in the middle. To provide a little clearance for the workpiece, cut out a couple of paper shims in the same shape as the fence piece. Put all the pieces together, drill some pilot holes, and screw the whole jig together with some bronze or brass screws **(B)**. By leaving it unglued, you can add or remove paper shims if the fence opening is too narrow or too wide. You could also cut and use fence pieces made to suit discs of other thicknesses.

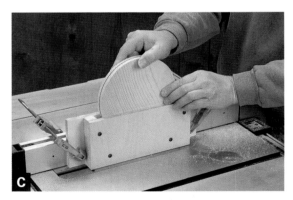

To use the fence for grooving and slicing, clamp it to the tablesaw's rip fence. With the sawblade fully retracted, adjust the rip fence so that the disc fence is positioned as desired over the top of the blade. Using a scrap disc, raise the spinning sawblade up into the disc set into the fence. Check and adjust cutting depth as desired, and then cut your "good" discs by lowering them onto the fence and blade, and rotating them slowly counter to the rotation of the blade to make the cut **(C)**.

This jig also works on the router table for edge-shaping operations. Cut a slot in the face of the clamping piece to provide access for the bit, and then clamp the jig to the router's regular fence **(D)**.

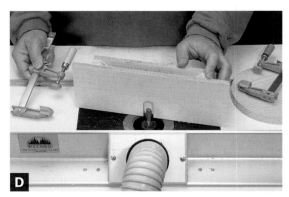

Arc-Cutting Fence

This fence jig makes it a breeze to saw the rails used for arch-topped cabinet doors. The fence is simply a square panel of MDF or plywood with a big hole cut out of it. The hole's edge creates a fence that guides the door rail as it's cut on the bandsaw **(A)**. To determine the hole size, clamp a cabinet rail on the bench and set a beam compass to draw the desired arc on the rail. The distance from the compass's pivot point to the far corners of the rail **(B)** equals the diameter of the fence hole.

> See *"Beam Compass"* on p. 52.

Cut the fence panel at least 3 in. larger on a side than the hole diameter. It's best to rout rather than saw the hole out, using a ¼-in.-dia. straight bit in a router fitted with a circle jig **(C)**. You don't need to rout a full circle—two-thirds of a circle is all that's needed for the jig to work. Rout it out in several successively deeper passes.

Sand the routed edge smooth, and then wax the edge with paraffin or an old candle to make the workpiece glide more easily. Clamp the fence jig atop your bandsaw table, positioning it so that the distance between the center point of the fence circle and the cutting edge of the blade equals the measurement you set the beam compass to when you drew the arc on the cabinet rail earlier. To cut an arc, set the top edge of the rail against the circle, and then slowly rotate the rail into the blade, moving clockwise **(D)**. Feed the rail slowly until the cut is complete **(E)**.

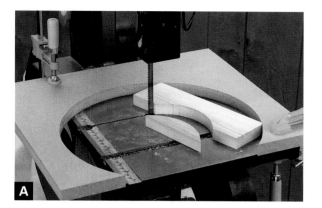

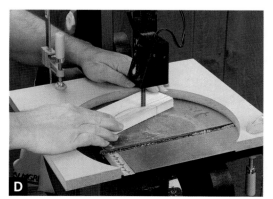

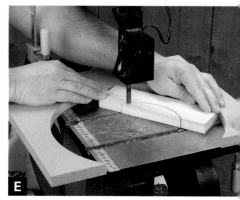

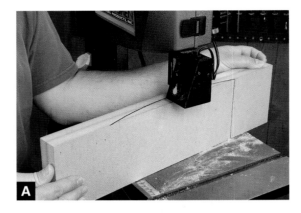

Curved-Part Shaping Fences

A shaper or router table fitted with a large bit is capable of creating many shapely parts, including oval moldings, curved panels, chair backs, table legs, and even curved railing sections. In each case, a curved fence is needed to support the stock as it's shaped or routed.

For raising the curved edges of bent coopered panels with a vertical panel raising bit or cutter, you need a tall fence to support the panel up on edge. You can saw a tall fence from thick lumber or from a lamination glued up from two or more layers of MDF or plywood. Using a bandsaw, cut out a concave segment with a radius that matches the panel's curve **(A)**. Wrap coarse sandpaper around a piece of the convex cutoff to serve as a sanding block to fair the sawn surface **(B)**. Glue and screw the tall fence to a plywood or MDF base that's large enough to allow clamping the fence to your shaper or router table **(C)**. Trace a slightly undersized outline of the panel raising bit/cutter onto the fence at the center of the lower edge, and then cut it out with a jigsaw or coping saw. Once the fence is secured in position over the bit, feed the panel past the bit, moving against its rotation **(D)**. It's best to complete the cut in several light passes, increasing the depth of cut each time.

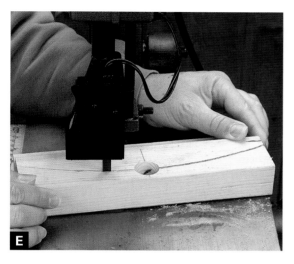

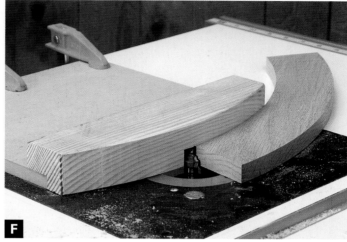

For grooving or shaping the convex or concave edges of curved wood rails and circular segments—say, for chair rockers or a spinning wheel—make a simple curved fence from a chunk of 2-in.-thick lumber wide enough and long enough to suit the workpiece. Using a bandsaw, cut one edge of the chunk to a curve that matches the radius of the work **(E)**, but before cutting the curve, bore a hole at its midpoint to accommodate the router bit or shaper cutter. Make a plywood or MDF base as wide as the fence is long, and attach it to the back edge of the fence to allow clamping to the router table or shaper **(F)**.

Have an even more complicated part to shape? Say, a section of spiral stair railing? You can create a compound curved fence to suit just about any shaping situation by combining two curved blocks. Using thick lumber, cut the lower fence block to match the curve on the bottom of the workpiece, boring a hole for the cutter as before. Cut another fence block, which matches the edge of the work, for guiding it past the cutter. Now shape one edge of the lower block so that it will mate with the other block **(G)**. Screw the two blocks together and mount them atop the router or shaper table by attaching them to a base or the machine's standard fence for clamping **(H)**.

Dowel-Splitting Guide

This bandsaw jig is designed to accurately slice dowels. You can use it to cut a kerf in the end of a dowel for using a wedged tenon, or to saw a dowel entirely in half lengthwise (see photo A on p. 75).

To make the jig, start by cutting a 2-in.-wide, 1⅝-in.-deep V-groove down the center face of a strip of 2x4 lumber at least a foot long, as described in the next photo-essay on the dowel-pointing guide. On the bandsaw, cut a 1-in.-wide notch, positioned 3 in. from one end of the V- block (**A**). Extend the notch just past the center of the V-groove. Crosscut a 3-in. length from the other end of the strip, flip it over, glue it atop the strip next to the notch, and then bandsaw a ¾-in.-long slot precisely at the center of the V-groove (**B**) to accept a kerf-alignment vane.

Make the vane by cutting a 3-in.-long section from an old bandsaw blade with the teeth ground off. Wrap both ends of the vane with tape until it fits snugly in the kerf (**C**). Secure it in the kerf by first drilling through the block and vane at each end and then pinning it in place with a couple of finish nails.

Glue the jig atop a square base cut from ⅛-in. or ¼-in.-thick hardboard. Clamp the jig to your band-saw table with the blade set into the notch just ahead of and perfectly parallel to the vane. Make a test cut on a scrap of dowel, and realign the jig if necessary. Finally, glue a wood strip (sized to fit your saw's miter-gauge slot) to the jig's base to facilitate positioning of the jig for future use (**D**).

[TIP] The dowel-splitting guide can also be used to cut a kerf in the ends of square tenons to accept diagonally placed locking wedges.

Dowel-Pointing Guide

If you have a stationary disc sander, this jig makes it a breeze to chamfer the ends of dowels and rods or to put a sharp point on them. The jig has two V-grooves: one for small-diameter dowels, and one for dowels up to 2½ in. in diameter.

Make the guide from a 10-in.- to 14-in.-long block of 2x4 lumber. On the tablesaw, cut two V-grooves into one face of the block: one ½ in. wide, and the other 1½ in. wide **(A)**. The next step is to miter the end of the block to the desired angle, which will depend upon how sharp a point or how steep a chamfer is desired. At the extremes, an angle of 45 degrees will yield a more obtuse tip or chamfer, while a 15-degree angle will create a very pointy end.

After cutting the angle, glue the block atop a ¼-in.-thick hardboard base. With a thin cardboard spacer sandwiched between the miter face and your sanding disc **(B)**, glue a wooden runner (sized to fit your sander's miter-gauge slot) to the underside of the base. This keeps the jig aligned while allowing it to slide back and forth to make use of the entire sanding disk. To protect your fingers, attach a small piece of thin hardboard to the base **(C)**.

To create a point or chamfer, simply place a dowel in the groove with the end pressed against the disc, and rotate it by hand. For more precise work, you can fit it with a wooden stop collar, which butts up against the end of the V-block to restrict the thrust of the dowel against the disc **(D)**.

[TIP] You can use the dowel-pointing guide to put a precise point on pencils with round shanks.

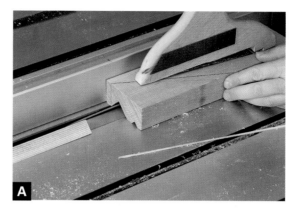

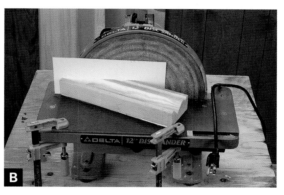

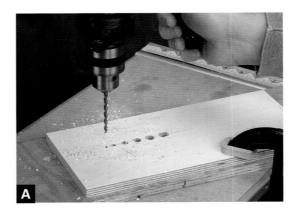

Dowel-Sizing Jig

This sanding jig quickly and accurately reduces the diameter of any dowel you can chuck in your drill press. Start by making a 6-in. x 12-in. baseplate for the jig from ½-in.-thick plywood or MDF. Bore a row of holes through the base, located approximately as shown (A). Include one hole for each desired size of dowel, spacing the holes at least ⅜ in. apart. Each hole should be ¹⁄₃₂ in. larger in diameter than the nominal dowel size. For example, drill a ⁹⁄₃₂-in.-dia. hole for ¼-in.-dia. dowels.

Cut the pivoting sanding arm from a piece of 2x4 lumber. Make it 1⁵⁄₁₆ in. wide x 9¾ in. long. (This particular arm is sized to accept a standard 3x21 sanding belt that has been slit and torn in half lengthwise.) Bandsaw the ends of the pivot arm to a half-round shape and then sand them smooth (B). Also cut a ¼-in x 1⁵⁄₁₆-in. x 8¼-in wood shim, and sand its ends to a taper as shown. The shim slips in between the belt and the back side of the pivot arm to keep the belt snug. Bore a hole through the base near the corner opposite the dowel holes, and mount the arm with a wood screw (C).

Chuck a dowel into the drill press with the end you want to size poking down through the appropriate-sized hole in the jig and extending down through the drill-press table. Now clamp the jig down, set the drill press to a low speed, and switch it on. Pivot the sanding arm against the dowel, using light pressure while feeding the dowel down through the hole to size for the desired length (D). Check the dowel diameter with calipers. As the sanding belt wears, rotate it on the pivot arm.

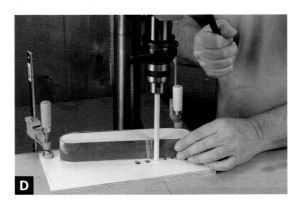

Sliding and Pivoting Jigs

WHETHER THEY'RE CALLED carriages, sleds, or carriers, jigs that slide are an important part of any woodworker's arsenal. Sliding jigs work with tablesaws, bandsaws, shapers, disc sanders, planers, and more to move a workpiece in a controlled fashion past a cutter, blade, or abrasive. They're useful for innumerable machining operations, from mitering stock and cutting joinery to planing warped boards. They're also used for sawing circular parts and complex shapes and for safely shaping short parts and narrow moldings.

A simple example of a sliding jig is the miter gauge that comes with every tablesaw. It holds the stock at a fixed angle and slides it past the sawblade in a straight line. In fact, a standard miter gauge serves as the basic "sliding hardware" for a number of jigs. For example, you can cut box joints, say, for constructing sturdy and attractive drawers for a chest or cabinet, by mounting a specially made plate with a reference pin to the head of a miter gauge, as seen in the photo on p. 101. You can create other sliding jigs by using a guide bar mounted to the a jig's base to slide in the miter gauge slots found on the majority of stationary woodshop machines.

Miter Slot Guided

Miter Gauge Jigs

Fence Guided

Table Guided

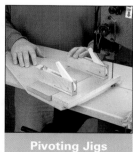

Pivoting Jigs

Guided by the tablesaw's miter gauge slot, this dovetail sled can be used in four different positions, for cutting out left- and right-hand sides of both pins and tails that form the joint.

Mounted to a handled jig that slides atop the tablesaw's rip fence, this tapering jig adjusts for a wide range of angles and can even cut out triangular parts.

The reversible dovetail sled, shown in the top photo at left, employs two guide bars for a controlled sliding motion that produces clean pin and tail cuts.

Miter gauge slots aren't the only means of moving a jig in a straight line. Fences, whether factory-made or custom-made, are also excellent guides for jigs that carry a workpiece in a straight line. A universal fence jig relies on a standard tablesaw rip fence for guidance. The device's tunnel-shaped body slips over the fence and accepts different jig setups, such as the tapering jig, shown in the bottom left photo.

A machine's table can also provide adequate guidance for a sliding jig. For example, a planer carriage slides in a straight line, relying only on the planer's table for guidance. Other sliding jigs don't slide at all but, rather, pivot to perform their task. Pivoting jigs are essential for cutting perfectly circular discs and rings as well as for rounding table corners and cutting precise arcs.

A circle-cutting jig fits on the bandsaw and pivots a square workpiece past the blade on an adjustable pin, allowing you to cut out perfect circular discs and rings.

Miter Sled

You can do a pretty good job of mitering small picture frame members using a regular miter gauge on the tablesaw. But a sliding miter sled not only allows you to cut longer, wider stock with less effort, but it produces perfect, tight-fitting corners every time.

Start by cutting a jig base from ¼-in.-thick hard-board or ½-in.-thick MDF or plywood that's at least 20 in. wide and 24 in. long—bigger if you're going to tackle really long, wide frame members. From straight-grained 1-in.- or 1½-in.-thick stock, cut two fence bars at least 18 in. long and a 3-in.- to 4-in.-wide cross rail that's one-quarter to one-third shorter than the width of the base. Miter both ends of the cross rail, and glue and screw it to the leading edge of the base **(A)**.

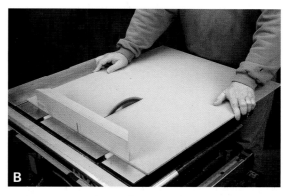

▶ See *"Installing Guide Bars"* on p. 21.

Mount the base atop a pair of miter guide bars with the cross rail perpendicular to them, and then carefully "cut in" the jig by sawing through the cross rail and base **(B)**. Stop the cut about halfway across the base, and unplug the saw with the blade raised. Use a framing square to position the two fences square to one another, and a miter square and straightedge to align the fences at 45 degrees relative to the sawblade **(C)**.

Glue and clamp the fences in place. Cut a triangular block from 2-in.-thick stock, and glue it at the junction of the fences to serve as a protective exit block.

Adding a strip of PSA sandpaper to each fence face helps keep the frame stock from slipping as it's cut **(D)**.

Crosscut Sleds

If you don't have a big radial-arm saw or panel saw, crosscutting long and wide workpieces requires large sliding jigs. To make square cross-cuts on thick planks and large panels, a twin-runner sled uses both of the saw's miter gauge slots to align the workpiece and guide it in a straight line past the sawblade (A).

Make the sled's base from ¼-in.-, ½-in.-, or even ¾-in.-thick MDF or plywood, sized as big or small as desired. Make it at least a few inches wider and at least half as long as the longest workpiece you wish to cut. Glue and nail a cross rail cut from 2x stock to the top edges of the base at both the front and back ends. Make certain that the rear member, which will serve as a fence, is perfectly straight, with a face that's square to the base.

Set a pair of shopmade or commercial miter guide bars into the miter gauge slots on your tablesaw, align the jig's fence to be perpendicular to the bars (B), and fasten the jig base to them. For safety, mount a protective exit block to the back of the fence. Hold the workpiece firmly against the fence as you saw, and support the end of extra-long work as necessary, so it doesn't tip the sled.

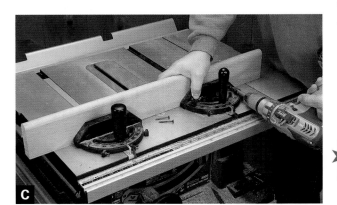

➤ See "Exit Guards" on p. 247.

The downside of the twin-rail sled is that the thickness of its base reduces the cutting capacity of the sawblade. For square crosscuts in really thick stock, you can make a "bottomless" cross-cut sled in either of two ways: The simplest method is to attach a pair of miter gauges to a single fence bar (C). Make the fence from stock

that's at least 1 in. thick and at least 1½ in. to 2 in. wider than your saw's maximum cutting depth. For safety, add an exit block. If you don't have a couple of miter gauges handy, make the jig **(D)** by mounting the fence to a pair of miter bars, adding a few glued-up layers of stock behind the fence to reinforce it and to act as a protective exit guide.

To build a sliding jig that handles miters as well as square crosscuts **(E)**, start by making an adjustable fence from several lengths of ½-in. thick MDF or plywood. The idea is to create a sandwich of two fence faces **(F)**, with shorter lengths of stock in between to create a pair of slots to accept ¼-in. carriage bolts. One will be a pivot bolt and the other will be a locking bolt. These allow adjustment and angling of the fence as well as back-and-forth movement for setting the end of the fence flush to the line of cut.

Make the base from ¼-in.-thick tempered hard-board, with a miter guide bar fastened to its underside so that one edge of the base is flush to the saw's line of cut. Drill a counter-bored hole in the middle of the base for the pivot bolt, and rout a slot for the locking bolt, as shown **(G)**. (If necessary, rout a wide shoulder into the slot to recess the head of the carriage bolt.) After set-ting the fence to the desired position, lock it down with a pair of threaded hand-knobs. A sin-gle miter bar fastened to the bottom of the base guides the jig on the tablesaw **(H)**.

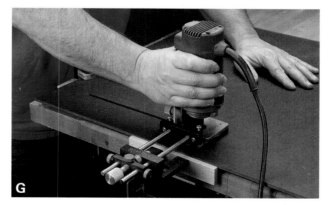

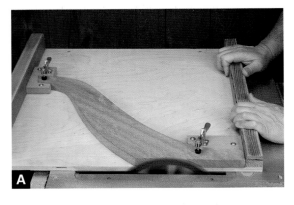

Part-Trimming Sled

A sliding sled jig for the tablesaw can also be put to good use for trimming all manner of circular, curved, and irregular parts that don't have a straight, true edge to butt up to a fence. This jig is terrific for trimming straight edges or cutting tenons and lap joints on curved legs and other parts that have been cut out of wide boards and panels laid out for maximum parts yield **(A)**.

Build the sled just like the twin-rail crosscutting sled described in the previous photo-essay on p. 94. The difference is that this sled is guided by only one miter guide bar, and one edge of its base is aligned flush with the saw's line of cut. Make the base from ½-in.-thick MDF or plywood, sizing it to accommodate your largest parts. Glue and screw a 1-in. or 1½-in.-thick cross rail to the front and back edges of the base to keep it flat. Position the jig atop the saw table so that one edge is parallel to the line of cut, while overhanging it by ⅛ in. to ¼ in. Fasten a miter guide bar to the underside of the base **(B)**. Then "cut in" the edge of the jig by running it past the blade **(C)**.

▶ See *"Crosscut Sleds"* on p. 94.

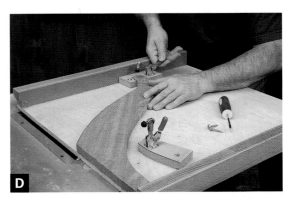

To accurately position the workpiece on the sled for trimming, cut several stop blocks, each shaped specifically to fit snugly against the edge of the work **(D)**. Unless it's safe to hand-hold the part, employ some kind of hold-down device, such as toggle clamps, to secure the work firmly in place. This is especially important near the line of cut, where a clamp serves to keep the work flat and your hands out of harm's way.

Bevel Sleds

It's possible to bevel the edge of a panel on a tablesaw by tilting its blade, but it's a whole lot easier to build a ramped jig that holds the workpiece at an angle relative to the sawblade, eliminating setting and re-setting of the blade angle.

To make a simple jig for beveling the edges of panels, planter box sides, etc., start by cutting and assembling a triangular, ramped "box" **(A)**. To determine the angle of the ramp, subtract the desired workpiece bevel angle from 90 degrees. For example, a workpiece bevel of 52 degrees subtracted from a 90-degree right angle yields a ramp angle of 38 degrees. Align the edge of the jig flush with and parallel to the saw's line of cut, and attach a miter guide bar to the underside. To align the workpiece, tack a straight wood fence to the top front edge of the jig. Screwing a toggle clamp to the fence helps keep the workpiece flat during beveling **(B)**.

You can also make a bevel jig that's very useful for cutting 45-degree corners for small boxes or drawers **(C)**. This jig has two tilted tables for making both right- and left-hand cuts. Make the base for the jig by fastening a cross rail to the back edge of a piece of ¼-in.-thick hardboard and attaching a pair of miter guide bars to the bottom that will guide the jig over the sawblade. Make each tilt table from ½-in.-thick plywood or MDF, gluing and nailing two rectangular tables—each with a mitered lower edge—to two supports shaped like right triangles **(D)**. Attach the tilt tables to the jig base with their lower edges just touching at the line of cut.

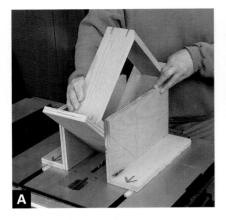

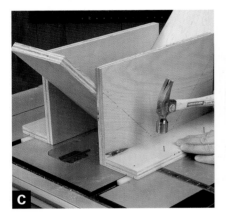

Spline Jig

Cutting exposed splines to reinforce the corners of a box or frame is easy with a sliding jig that supports an assembly at a 45-degree angle as it passes over the tablesaw blade **(A)**. Make the jig by first constructing two L-shaped side supports from ¾-in.-thick plywood **(B)**. Cut two rectangular tilt tables, each at least half again as wide as the widest box side you plan to cut. Bevel the lower edge of each table at a 45-degree angle. Glue and nail the tables to the side supports to form a V, with each beveled edge facing downward and flush with the bottom edge of the support.

Use a miter gauge to accurately align the tables. Lay a pair of wood or plastic miter guide bars into your saw's miter slots and set the spline jig on top of them. Position the jig so the side support is parallel with the slot, and center the jig side-to-side over the sawblade. Screw or nail down through the side supports to fasten the jig to the rails **(C)**. To cut a pair of splines on each of the corners of a box, set the box into the jig with one edge against a side support, and make the cuts with the sawblade height adjusted to cut the slot to the desired depth **(D)**. After cutting all four corners of the box in this fashion, set its other edge against the opposite side support and cut the other four spline slots. You can change spline spacing on a box or cut splines in picture frame corners by using a spacer with the jig to align the workpiece relative to the blade as desired **(E)**.

Dovetail Sled

Hand-cut dovetail joints may be a hallmark of fine craftsmanship, but that doesn't make them any less tedious to cut. If you want the look of hand-cut joints with a lot less handwork, then try this tablesaw method, which uses a sliding dovetail sled to cut 90 percent of each joint. The drawing at right shows the basic dimensions and layout of this sled, which you can build from ½-in.-thick MDF or plywood.

Cut the pair of pin fences first, beveling the inside-facing end of each at 92 degrees. Glue and nail two triangular braces to each pin fence and then attach them to the baseplate, positioning them as shown **(A)**. Next, cut two pairs of 8-degree wedges from ¾-in.-thick stock, and glue them together to form a wide ramp that angles the workpiece for the tail cuts. Glue and nail each ramp to the edge of the base **(B)**. Butt the tail fence against the inside face of the ramps, and fasten it in place using three more triangular braces to make it stronger and keep it perpendicular to the base **(C)**.

(Text continued on p. 100.)

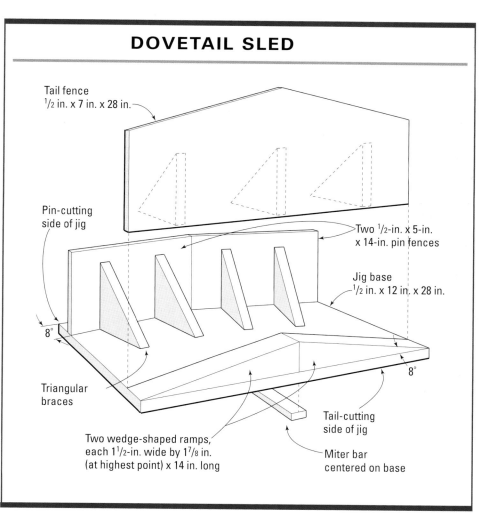

DOVETAIL SLED

Tail fence
½ in. x 7 in. x 28 in.

Pin-cutting side of jig

Two ½-in. x 5-in. x 14-in. pin fences

Jig base
½ in. x 12 in. x 28 in.

8°

Triangular braces

8°

Two wedge-shaped ramps, each 1½-in. wide by 1⅞ in. (at highest point) x 14 in. long

Tail-cutting side of jig

Miter bar centered on base

A

B

C

Attach a single miter guide bar to the underside of the base, centering it and making sure it's dead square to the base's long edges. Now set the sled's bar into one of the tablesaw's miter slots and carefully "cut in" one of the pin fences (D). Then set the bar into the saw's other slot and cut in the opposite pin fence. Next, flip the sled around and cut in each tail fence ramp, setting the jig into one miter slot and then the other, as with the pin fences. To make the jig safer to use, glue a 2x4 exit block to the back of each fence, directly over the saw kerfs you cut. To prevent the blade from cutting through the exit block, use a miter slot stop during all the cutting steps listed next.

► See "*Travel Stops*" on page 198.

Start by marking out the size and spacing of the dovetail pins and then cutting them in sequence: Cut the waste portion on one side of each pin, holding the stock firmly against the first pin fence (E), making sure the cuts angle in the proper direction. With the jig set into the saw's other miter slot, use the second pin fence to cut the other side of each pin. You can either chop the rest of the pin waste away with a chisel, or remove the waste with multiple saw cuts (F).

As in hand-cutting dovetails, transfer the outline of the completed pins to the tail portions of the dovetail joint. Then reverse the jig and use the tail-cutting ramps and fence to remove the waste at each edge of each tail (G). To save time, you can remove the waste to finish the outer two tails on each workpiece by cutting them with a miter gauge on the tablesaw (H), but you'll have to chop the waste for the remaining tails the old-fashioned way: with a chisel and mallet.

Box Joints

You can cut very clean and tight-fitting box joints on the tablesaw with a dado blade and a sliding box-joint jig. The jig consists of an 8-in. by 18-in. vertical fence, cut from ¾-in.-thick plywood, attached to a miter gauge. A short length of T-track screwed onto the back of the fence provides attachment to the gauge head with a pair of ¼-in. bolts **(A)**. This allows a short range of side-to-side adjustment that's necessary for fitting the joints.

To make the jig, install a dado head that's set to cut to the width of the desired box joint fingers (in this example, ½ in.). Raise the dado head about the same distance as its width, and cut a notch into the fence's lower edge at about center. Use a dense hardwood like maple to make a 1½-in.-long pin to fit the notch exactly **(B)**. Glue the pin into the notch so it protrudes from the front of the fence. Now shift the fence to the right until the distance between the pin and the edge of the dado blade exactly equals the pin width.

To cut the first half of a joint, butt the edge of the work against the pin, holding or clamping the work firmly against the fence while sliding the jig smoothly over the blade **(C)**. For all subsequent cuts, register the previously cut notch on the pin **(D)** until you've cut notches all the way across. To cut the joint's other half, space the stock a pin-width away from the blade for the first notch, and then cut subsequent notches as before. Check the fit of the joint: Too loose? Shift the fence slightly to the right. Too tight? Move it to the left.

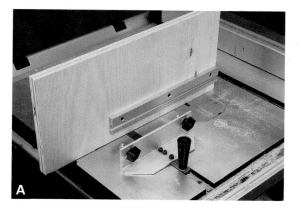

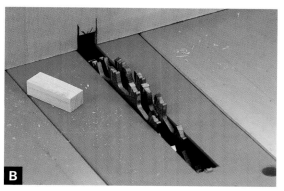

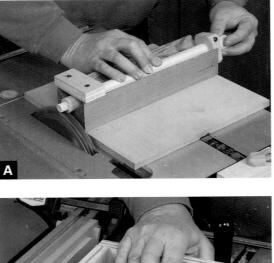

A

B

C

Round Tenons

Round tenons can create strong connections between parts in furniture and are a good choice when, say, joining the leg of a chair or barstool to a wooden seat. Tenoning dowels, turned spindles, and other round stock on the tablesaw is a breeze using a sliding V-block jig **(A)**. Cut the block from a length of 4x4 and mount it to a square plywood base with its end flush with the edge. Fasten a miter guide bar to the base, positioning it so the V-block is perpendicular to the bar with its end at the line of cut. Slide a stop collar onto the workpiece (see photo B on p. 197), set it into the V, and slide the jig forward until the work is centered over a dado blade whose height is set to create the desired tenon diameter.

You can cut round tenons on square stock using the jig shown **(B)**. Measure the work diagonally, and bore holes through two ¾-in.-thick plywood squares on the drill press **(C)**. Mount the squares to a bar that's 1 in. to 2 in. shorter than the workpiece, and screw the assembly to a miter gauge so that one end is ¼ in. from the line of cut, clamping a stop strip to the saw's rip fence **(D)**. Slip the workpiece into the jig (adding tape to the corners if necessary to correct a loose fit), and center the work over the blade, using a stop block in the miter slot to prevent forward motion of the jig. Rotate the work until the tenon is cut. The position of the fence stop strip determines final tenon length.

D

Small-Part Cutting

Cutting very short and/or thin workpieces to length on the tablesaw usually places your hands way too close to the blade—a prelude to possible disaster. This small-parts cutting jig **(A)** mounts to a regular miter gauge and uses old-fashioned wooden clothespins to hold tiny parts firmly and safely as they are crosscut to length.

Start by taking four clothespins apart, and trimming the head off one-half of each pin, using a small razor or dovetail saw **(B)**. Cut the head off just beyond the groove for the clothespin's wire spring and then reassemble the clothespin. Make the sliding base and fence for the jig by gluing together a 4½-in.-wide strip of ¼-in.-thick tempered hardboard to the edge of a ¾-in. x 2-in. strip of wood long enough to roughly center the base over your tablesaw's line of cut.

Mount the assembly to your miter gauge with a pair of screws, and then "cut in" the jig with the sawblade raised only about ¾ in. **(C)**. Set the trimmed and reassembled clothespins onto the jig's base, with two of them closely flanking the saw kerf, and then use masking tape to hold them in place. Apply a few drops of medium- or thick-consistency cyanoacrylate glue to cement the lower half of each clothespin in place where it meets the base **(D)**. Finally, apply a couple of strips of PSA-backed sandpaper to the base and fence to help prevent small parts from moving as they are clamped down by the clothespins and cut.

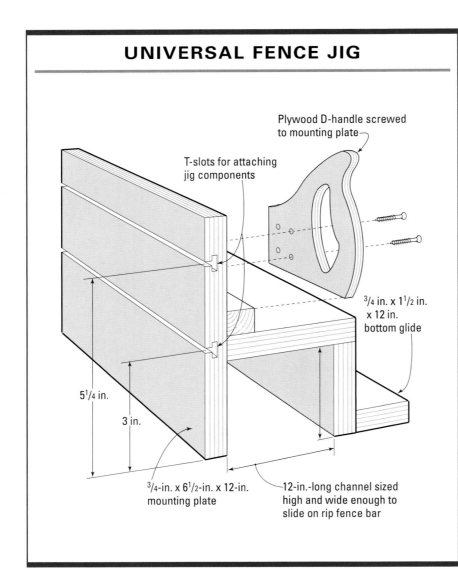

UNIVERSAL FENCE JIG

Plywood D-handle screwed to mounting plate

T-slots for attaching jig components

$^3/_4$ in. x 1$^1/_2$ in. x 12 in. bottom glide

5$^1/_4$ in.

3 in.

$^3/_4$-in. x 6$^1/_2$-in. x 12-in. mounting plate

12-in.-long channel sized high and wide enough to slide on rip fence bar

A

B

Universal Fence Jig

Unfortunately, a miter-slot-guided tablesaw jig requires a complicated mechanism to adjust the distance between the work and the sawblade. This universal fence jig sidesteps that problem by working with a standard rip fence, using the same adjustment mechanism you use for setting rip cuts. The jig is "universal" because it accepts different fences, tables, stops, and accessories, so it's useful for cutting tenons, raised panels, splines, taper cuts (see the bottom left photo on p. 92), and more.

Most parts for the jig are cut from plywood to the dimensions shown in the drawing at left. The jig's inverted U-shaped channel is sized to fit over your saw's rip fence. Size the fit so there's about $^1/_{16}$ in. of side-to-side clearance between fence and channel; you'll shim it snug later. Before assembly, cut two T-slots into the jig's mounting plate, first dadoing two grooves and then finishing up with a T-slot bit in a router or laminate trimmer **(A)**.

Once the basic jig is glued and nailed together, cut a handle from $^3/_4$-in.-thick plywood, tracing the shape of your favorite handsaw's handle. Screw it on the upper corner of the vertical plate **(B)** for better control of the jig. (The screws allow removal in case the handle impedes an operation.) Now apply a few lengths of sticky-backed UHMW or nylon tape inside the channel to shim it to fit snugly over the fence **(C)**, applying more than one layer, if necessary. Also stick the tape on the lower edges of the jig so it glides smoothly on the saw table.

The jig is very adept at cutting tenons. The vertical plate provides a sturdy mount for a tall fence that'll handle long workpieces. Make the tenon fence by gluing a 10-in.-long straight piece of $^3/_4$-in.-thick stock to a 1$^1/_4$-in. x 10-in. plywood mounting strip that's been drilled to accept two short T-bolts that secure it in place **(D)**. Add a small plywood flange at the bottom of the fence to serve as a hand rest and blade guard. Use a square to align the fence with the saw table, and secure the tenon fence with small hand knobs. If you wish, add an eccentric clamp to hold long stock firmly during tenoning **(E)**.

▶ See *"Cam Clamps"* on p. 215.

A bevel table lets you saw simple raised-panel edges without tilting your sawblade. Cut out a table that's a little bigger than your biggest panel, and tack stop strips to two adjacent edges **(F)**. After cutting the two triangular supports that will hold the table at the desired tilt angle, fasten them to the back side of the table. Attach a narrow plywood strip to each triangle after each has been been drilled with two holes for the T-bolts that mount the bevel table to the fence jig **(G)**.

By making a wooden fence with the same dimensions as your rip fence and clamping it to your router table, you can use the universal fence jig for a variety of shaping operations. For example, you can mount a spline-cutting setup to rout slots for dovetail splines in the corners of a box. Similar in construction to the spline-cutting sled, this setup involves a pair of ½-in.-thick plywood tables mounted in a V configuration to a single hardboard plate that has been drilled in three places for the T-bolts that mount it to the universal jig **(H)**. The V tables position the box as the jig guides each corner past a dovetail bit **(I)**.

▶ See *"Spline Jig"* on p. 98.

[TIP] By slotting the holes that mount the tenon fence to the universal jig, you can tilt it for cutting angled tenons.

C

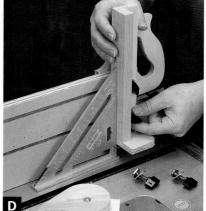
D

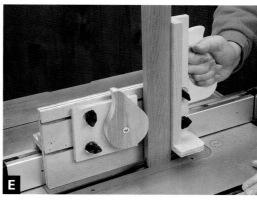
E

F

G

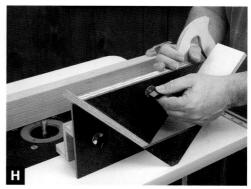
H

I

Edge Trimming

Trimming the edges of rough, irregular planks can be tricky. A chainsawn plank may be too heavy to run over the jointer, yet you can't safely rip the plank without a straight edge to run against your saw fence. Fortunately, there's a very simple solution: Temporarily tack a guide strip of wood over one of the rough edges, and use the strip's edge against the rip fence to guide the plank past the sawblade to trim its opposite edge straight **(A)**. Fasten the strip atop the plank with brass screws, driving several of them along the plank's length **(B)**. Depending on just how rough and wavy the surface of the board is, you may have to use spacers or wedges to create a firm connection between the guide strip and the plank.

Another tricky tablesaw trimming task (try saying *that* 10 times quickly) is shaving just a skosh off the edge of a thin wood strip, because the strip may bind between the blade and fence and scorch or even kick back. A narrow-parts jig safely holds the workpiece and trims its edge cleanly by feeding it past a sawblade that's partially buried in an auxiliary fence face.

Cut the jig's base from MDF or plywood that's at least as thick and long as the workpiece. Nail a stop strip to the back edge of the base **(C)** and a ¼-in.-thick hardboard top cover piece to create a "pocket" to capture the workpiece. The cover is wide enough to overhang the workpiece slightly and bears against the fence face to guide the trim cut past the partially buried sawblade **(D)**. The sawblade is raised high enough to cut the part, but not the hardboard cover. Adding a handle to the jig makes it easier and safer to control.

Tapering Jigs

Taper cuts are necessary for creating a variety of parts, from simple wedges for a clamping jig to elegant legs for a demilune table. The tablesaw and a taper jig are ideal partners for making these parts safely and accurately. The most basic taper jig need be little more than a rectangular jig stick with an angled notch cut in it. As we'll see, the notch holds the workpiece at an angle, while the stick's straight edge bears against the rip fence as one edge of the work is cut to a taper. This is great for cutting a wedge or a right triangle, but tapered parts like chair and table legs are often tapered on all four sides. The same notch that positions the work when cutting the taper on one edge won't produce the same taper when the work is flipped over for cutting the opposite edge.

Four-sided tapers (as well as two-sided tapers, such as a guitar's fingerboard) require a jig stick with two notches **(A)**: One notch fits the work's uncut edge, and the second positions the previously tapered edge. When tapering a four-sided square leg, use the first notch to cut two adjacent sides of the part **(B)**. To keep cuts square, place an uncut side of the leg face-down for the second cut. Flip the jig over and use the second notch to taper the remaining two sides of the leg **(C)**. This time, to keep the cuts square, you must tape the scrap pieces from the first cuts back onto their tapered sides before cutting the remaining two.

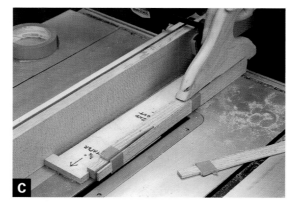

ADJUSTABLE TAPER JIG

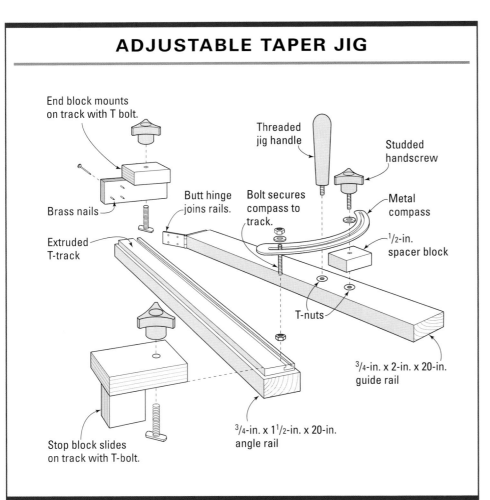

End block mounts on track with T bolt.

Threaded jig handle

Studded handscrew

Brass nails

Butt hinge joins rails.

Bolt secures compass to track.

Metal compass

Extruded T-track

½-in. spacer block

T-nuts

¾-in. x 2-in. x 20-in. guide rail

Stop block slides on track with T-bolt.

¾-in. x 1½-in. x 20-in. angle rail

While it's more complicated to make, an adjustable taper jig is much more versatile. You simply set the jig to the required taper angle instead of cutting out a new jig stick for each job. Following the drawing at left, cut the jig's two rails from straight-grained stock. The guide rail rides against the saw's rip fence, while the angle rail adjusts to position the workpiece for tapering. Clamp the two rails together and attach a butt hinge to connect them at one end **(D)**. Drive the screws into pilot holes so you don't split the wood.

The angle between the rails is adjusted and set using a commonly available curved metal compass **(E)** from a chest lid support. One end of the compass pivots on a ¼-in. bolt that's fastened to a length of extruded aluminum T-track screwed atop the fence rail **(F)**. The compass slot fits onto a threaded handscrew driven into one of two T-nuts installed in holes drilled in the fence rail. Repositioning the pivot bolt in the T-slot and moving the compass and handscrew from one T-nut to the other let you change the jig's range of adjustment. A threaded handle screws into the unused T-nut to lend better control of the jig.

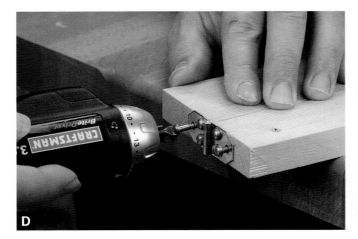

D

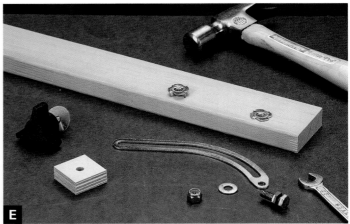

E

In addition to anchoring the pivoting compass, the T-track on the fence rail also serves as a mount for two pinch blocks that secure the workpiece to the jig during tapering. The end block is fixed at the hinge end of the jig. The stop block slides along the length of the T-track and is adjusted to suit the length of the workpiece. Small brass nails driven through the end stop (**G**) serve to grasp one end of the workpiece. A layer of PSA sandpaper applied to the stop block helps grasp the part's other end.

Before using the jig, set the angle of the rail to the proper taper, either with a bevel gauge or by measuring the distance between sawblade and the rail at both ends. With the jig's end block butted against a scrap block of wood clamped to your bench or saw table, hammer the workpiece so it engages the nail tips or the end block (**H**). Slide the stop block up until it firmly contacts the other end of the workpiece, and lock the block in place.

Now you're ready to cut the taper by sliding the jig's guide rail along the rip fence (**I**). Use the jig's handle and, if necessary, a push stick to keep the jig pressed firmly against the rip fence during the cut. Alternately, you can mount the taper jig to the universal fence jig, as shown in the bottom left photo on p. 92, to prevent the taper jig from drifting away from the rip fence during cutting.

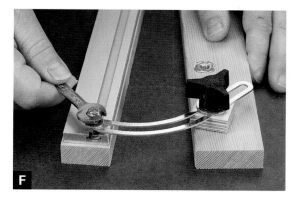

F

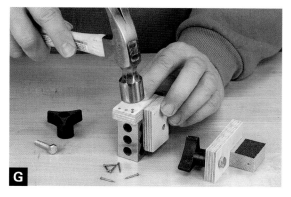

G

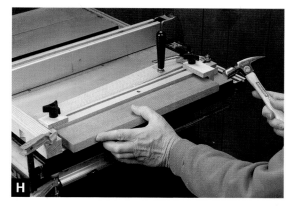

H

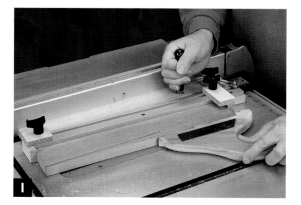

I

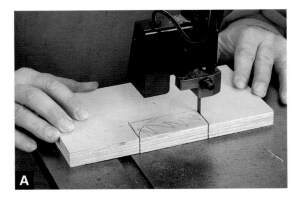

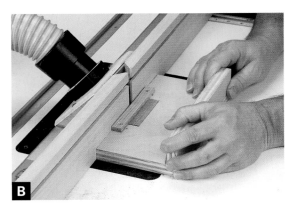

Shaping Jigs

This essay contains several sliding carriage jigs designed to help you perform a variety of shaping and routing operations more safely and successfully. The first jig provides a simple means of handling parts that are otherwise too small to shape safely. Start by making a plywood carrier board at least a foot long, 4 in. to 5 in. wide, and as thick as the part. Trace the part on one long edge of the carrier and saw along the outline **(A)**. Screw a wood strip over the top of the notched area, and add another strip of wood to the carrier board to serve as a handle **(B)**. Set the part into the notch, and slide the entire jig along your shaper/router table fence to make the cut. (If the part fits loosely in the notch, use tape or a thin shim to snug up the fit.)

Sometimes the problem isn't that a part is too small to shape, but that the edge you need to shape is too short. A small-parts carriage holds a part of practically any shape securely **(C)** and provides a guide that bears against the shaper/router table fence for safe and easy shaping. First, cut a base for the carriage from hardboard or plywood that's several inches longer and wider than the workpiece. Depending on the shape of the part, add whatever stop blocks and positioning strips are necessary to align the part with one long edge of the base. If you have lots of identical parts to shape, you'll save time and trouble by using a toggle clamp fastened atop one of the positioning blocks to secure the part to the base. Fit a pair of handles to the base's other edge **(D)**, and you're ready to go.

A part may be long enough to shape safely, but too narrow and/or thin to handle easily, such as in shaping "sticking" (the bars that separate the panes of glass) for a French-style door. Normally, one edge of the sticking is shaped on a wide piece of stock, and then the piece is ripped to width and shaped on its other edge. A narrow parts carriage **(E)**, holds the sticking firmly during the second operation. The jig consists of a base, a fence strip that supports the previously shaped edge of the part, stop blocks that cradle the part at both ends, and a pair of handles **(F)**. A clamping strip screws over the top of the jig to hold the workpiece securely in place during shaping.

The final jig in this essay is made to help when you're shaping a round part on the router table or shaper. A pair of discs screwed to the ends of the workpiece are the key to this operation **(G)**. The assembly is pressed against the router table fence and rotated and moved along as it's cut, a little at a time, by a long spiral bit **(H)**. A pair of stops clamped to the fence limit the side-to-side travel. After every pass, the depth of cut is increased by repositioning the fence no more than ⅛ in. per pass until the final shape is completed. Use two discs with matching diameters for a cylindrical form, or different-sized discs for a tapered form. Elliptical discs will create oval-shaped forms.

> ⚠ **WARNING** Standard bits cut too aggressively for this rounding. Always use a spiral-fluted bit protected by a plastic guard attached to the fence.

E

F

G

H

Planer Carriages

Jigs for the planer can be lifesavers, especially for straightening crooked stock. A carrier board takes only minutes to make but provides a sure way to flatten one side of a cupped and/or twisted board, so it can be planed true and to final thickness.

You can make a carrier board from ¾-in.-thick MDF or flat plywood. (Avoid construction-grade plywood, due to voids and inconsistent thickness.) Make the carrier a few inches longer than your workpiece and a little narrower than your planer's width capacity. Screw or nail a thin wood strip across the back edge of the carrier **(A)** as a work stop (use *only* brass hardware in any planing jig, in case the knives accidentally make contact).

Set the workpiece on the carrier with its end against the stop and the concave side facing down. Use whatever shims, strips, spacers, or wedges are necessary to keep the board from rocking and to prevent pressure from the planer's feed rollers from distorting it. Strips that run entirely under the board for support can be simply taped in place **(B)**. Wedges and shims near the edges should be screwed down. To secure the board during planing, you can drive counter-sunk brass screws through the board at the corners, or screw on a couple of stop blocks with small protruding nails **(C)** that grip the board's edges. Feed the board and carrier through the planer **(D)**, taking light passes until the entire board's surface is planed flat and true. Then remove it from the carrier and plane the other side as usual.

You can make a similar planer carriage to taper the thickness of a board along its entire length. To the bottom of the basic carrier board described above, glue on several wedge-shaped pieces cut to the desired taper angle **(E)**. Set the workpiece on top of the carrier, nearly flush with the jig's higher end, and screw stop strips to the sled at both ends and at the sides of the workpiece to secure it during planing **(F)**. Set the planer to take only a light cut on the first pass, and then keep reducing the planing thickness until the entire surface of the work is cut, leaving you with a perfectly tapered part **(G)**.

Another kind of carriage jig for the planer allows you to straighten and clean up surfaces and edges on parts that are either too short or too irregular to plane on a jointer. For the jig's base, cut a strip of ¾-in.-thick MDF or plywood that's 8 in. to 10 in. wide and as long as you wish. Cut two more strips 3½ in. to 5 in. wide, depending on your planer's maximum depth capacity. Clamp the strips together, and drill several ¼-in.-dia. holes along the center to accept carriage bolts. Fasten the long edge of one of the strips perpendicular to the base, about an inch from the edge. Place the workpiece against the strip with appropriate spacer blocks or wedges underneath it so that the edge you wish to plane is parallel to and just proud of the top edge of the strip **(H)**. Install a pair of bolts to sandwich the other drilled strip against the work, thus clamping it firmly. Pass the entire jig through the planer to take the cut **(I)**.

[TIP] To bevel a board's entire thickness width-wise, mount the board to a carriage that tilts the board side to side.

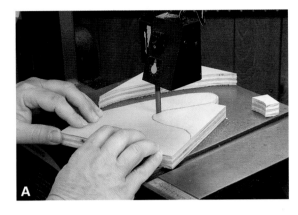

Small-Part Routing

Small workpieces usually don't mix well with power tool operations—especially routing and shaping. However, you can safely hold most small parts during light shaping operations on a router table with this double-handled jig. Saw each of the jig's two handles from a 7-in.-square piece of ¾-in.-thick plywood, giving them an offset shape **(A)**. The handles are connected by a pair of 12-in.-long rails, cut from ¾-in. x 1-in. solid wood. Nail and glue the ends of the rails atop two 1½-in. x 3½-in. spacer blocks cut from stock as thick as the thickest workpiece you plan to shape **(B)**.

Add a 2-in.-wide wood block flush with the outside edge of each spacer block to fill in the space between the rails. Drive a few brass nails or escutcheon pins through the middle of each rail so that their points project about ⅟₁₆ in. for securing the workpiece during routing. Drill and drive three screws through the plywood handles to mount them to the ends of the rail assembly **(C)**.

Set the workpiece under the rails to drive the nail points into the workpiece with a mallet. If the work is thinner than the jig's spacer blocks, or you don't want to mark up the work with the nail points, use a shim block and double-stick tape to mount the work under the rails. You can shape the edges of the work using practically any small piloted bit chucked into the router table **(D)**. To plow grooves with an unpiloted bit, guide the edges of the two handle bases against a router table fence.

Complex Shape Sawing

The bandsaw is great for sawing all manner of shapely forms—as long as the bottom surface of the workpiece sits flat on the saw table. But if you want to saw shapes on parts with curved or irregular bottom surfaces, you'll need a jig for sawing complex shapes.

The workpiece clamped to the box-like jig is oriented as desired and is firmly supported during the cut. Make the jig from several ¾-in.-thick plywood pieces glued and nailed together to form a five-sided box (**A**). The box can be as large as you wish, but it must be short enough to fit under your saw's guides with the work clamped in place. Set two of the box sides back from the edges of the top to provide clearance for small clamps that attach the workpiece or work pattern to the jig. The sixth side of the box is left open, so that a screw (or screws) can easily be installed through the opposite side to mount workpieces vertically.

The jig can be used either of two ways. To saw the profile directly, clamp the part (for example, a bracket foot for a chest, cabinet, or sofa) directly to the top of the jig with a couple of C-clamps (**B**). Then saw along the line marked directly on the part. In cases where it's impractical to draw the profile directly on the part (say, a disc), screw or clamp the part to the vertical front of the jig. Make a pattern for the desired cut from a piece of scrap ¼-in.-thick plywood or hardboard, and clamp it atop the jig, aligned as necessary above the workpiece (**C**). Follow the lines on the pattern to make your cut (**D**).

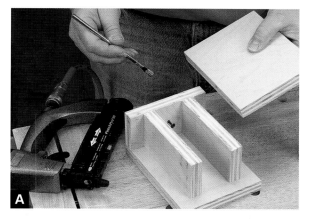

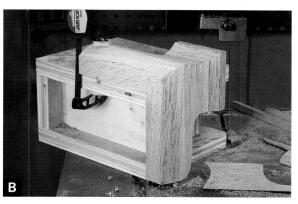

A

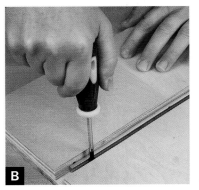

B

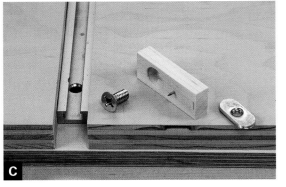

C

D

E

Circle Jigs

Cutting out perfectly circular wood discs is a breeze when you use a circle jig and a bandsaw. There's no layout with a compass or trying to follow a curved line with the saw. This jig mounts a wood square on a pivot point and rotates it past the blade to cut a perfect wheel, disc, or even a circular tabletop.

Make the base of the jig by gluing up three pieces of ¾-in.-thick plywood. The bottom is sized as you wish; a bigger base will cut bigger circles. The two strips glued on top of the base are narrow enough to leave a gap between them **(A)**, forming a groove wide enough for a T-track, which attaches with short wood screws **(B)**. Make a pivot block from a 2-in.-long strip of maple or oak. It should fit the groove snugly without binding, and its top should be flush with the top of the base **(C)**. For the pivot point, drive a small nail through a pilot hole so it protrudes about ³⁄₃₂ in. Drill a countersunk hole for a ¼-in. flat-head machine screw, and use a T-nut to mount the block to the track.

With the jig on your bandsaw table, position the base so it's just touching the side of the saw-blade. Mark the position of the miter slot on both edges of the base **(D)**, and mount a miter guide bar in this spot on the bottom of the jig with screws. With the jig in the miter slot, line up the T-track's centerline with the blade teeth. Attach a stop bar **(E)** to limit the jig's forward sliding motion at this point.

Set the radius of the desired circle by setting the pivot block's position, measuring between the nail point and the blade **(F)**. (Alternately, mount a sticky-backed measuring tape to the base.) Poke a small hole in the center of the square workpiece with an awl, and then slip it over the pivot nail and pound it on with a mallet. With the saw on, slide the jig forward until you hit the stop and the blade enters the edge of the workpiece. Slowly rotate the work clockwise until the cut is complete **(G)**. You can also saw a ring out of a disc by resetting the pivot to a smaller diameter and cutting the disc again, as shown in the bottom right photo on p. 92. (The resulting saw kerf usually glues up cleanly, so you can hardly see it.)

Here's a bonus: You can also use a circle jig to power-sand the edge of a disc perfectly smooth. Unscrew the jig's miter bar and set it in the stationary sander's miter slot. Refasten it to the jig's base so that the base's edge is nearly flush with the sanding disc or belt. Make a new sliding pivot block (without a screw hole), and a separate stop block that fastens to the track with a screw or T-nut **(H)**. Set the circle size by fixing the stop block where it will stop the forward motion of the pivot block at the desired distance. Clamp the jig base to the sander table, pound the pivot block onto the disc, and slide the block forward in the base, rotating the disc as it contacts the spinning abrasive **(I)**. When the pivot block contacts the stop block, the disc will have reached the desired size. For more effective sanding, use different areas of the sanding disc by repositioning and reclamping the jig in the miter slot.

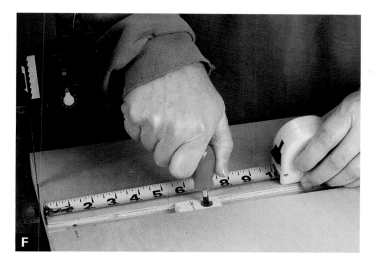

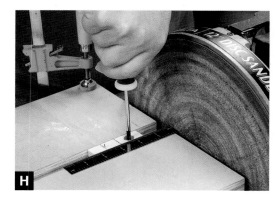

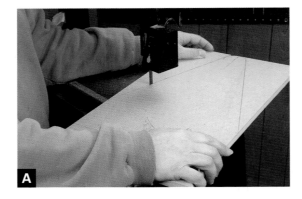

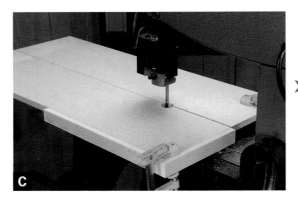

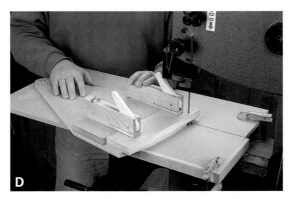

Arc-Sawing

An arc-sawing jig works with a bandsaw to rapidly cut curved parts, such as arched door rails or concave coopered panel forms. Cut the jig's pivot arm from a piece of ½-in.-thick MDF or plywood, making it an inch wider than the workpiece and 1 to 2 in. longer than the desired cut radius. On the bandsaw, taper the edges of the arm back towards a centered pivot point drilled an inch or so from the end **(A)**. The workpiece is positioned and fastened to the large end of the pivot arm with a stop strip and clamp assembly **(B)**.

Using stock that's about the same thickness as the workpiece, cut the 3-in. to 5-in.-wide stop strip as long as the pivot arm is wide. Screw the stop in place, locating it so the workpiece is flush with the end of the arm. Drive a pair of screws partially into the base to serve as end stops to keep the workpiece from sliding as it's cut. Two cam clamps attached with carriage bolts through counter-bored holes in the base secure the work. (Toggle clamps allow more clearance if saw guides interfere.)

➤ See *"Cam Clamps"* on p. 215.

Make the jig's MDF base long enough for the pivot arm and wide enough to support its full range of motion. Draw a lengthwise centerline on the base, and bore holes for the blade and pivot arm screw. Also cut a blade access kerf to the hole. Position the base with the blade teeth flush to the centerline **(C)** and clamp it in place, attaching two stop strips to the edges to relocate the base for future use. Clamp the work to the arm and smoothly swing it past the blade for a clean cut **(D)**.

Corner Radius Jig

To cut perfectly rounded corners on tabletops or stool seats without the fuss of marking them out and cutting them freehand, try this corner radius jig. Made from ½-in. or ¾-in.-thick MDF or plywood, the jig consists of a 12-in. x 15-in. base and an 11 in.-sq. pivoting plate **(A)**. On one corner of the pivot plate, draw an arc with a compass set to the desired corner radius, and then at the compass's pivot point, drill a counter-bored hole for a screw post.

Glue a pair of stop strips to the edge of the pivot plate adjacent to the radius corner.

Near the center of the baseplate, bore a hole for the bandsaw blade, and then cut a kerf to slip the blade through. Mark lines to show where the blade teeth will line up with the base hole, and then drill a hole for the pivot plate screw post that's aligned with the marks. (The distance from the hole to the blade teeth equals the corner radius.) After attaching the plate to the base with the screw post, screw two stop blocks to the base **(B)** to limit the pivot plate's range of motion to 90 degrees.

Clamp the jig to the bandsaw, aligning the teeth with the marks made earlier, then "cut in" the corner of the pivot plate **(C)**. If the blade cuts more than a rounded corner off the plate, reposition the stop blocks on the base. To use the jig, rotate the plate counterclockwise until it's against the stop block, set a corner of your workpiece against the stop strips on the plate, and rotate the plate/workpiece through the cut **(D)**. Repeat for each of the other corners.

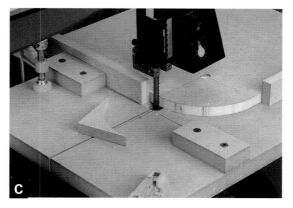

A

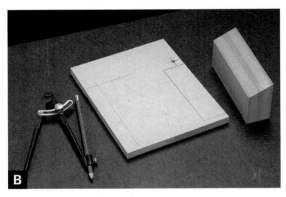

B

C

D

Compound Curve Sawing

Believe it or not, you can cut a complex shape, such as a helix-like spiral section of staircase hand rail, with a simple pair of cuts on the band-saw **(A)**. Make the jig's ½-in.- or ¾-in.-thick pivot plate first, cutting out a rectangle that's a few inches longer and wider than the workpiece. To support the workpiece at the desired angle, cut a tilt block that's at least 1 in. to 2 in. thick, with its top edge matching the desired angle of the workpiece. Mark both the tilt block's position and where the lower edge of the work rests **(B)**. Using a pencil compass, mark the two radii that define the outer and inner cutlines, positioning the compass's point slightly to the left of the tilt block. Mount the tilt block to the plate, and drill the ¼-in.-dia. pivot hole at the compass point.

Make a base for the jig **(C)** just like the one for the corner radius jig described in the previous photo-essay, adding marks that indicate where the blade teeth pass through the base. Drill two ¼-in.-dia. holes (counter-bored for carriage bolts) at distances that equal the two desired cutting radii, measured from the blade marks. Attach the workpiece to both the tilt block and pivot plate with a few brass screws. Set the pivot bolt into the larger-radius base hole, and clamp the base to the saw table with the blade teeth properly aligned to your marks. Pivot the work smoothly past the blade to cut the part's outer edge **(D)**. Move the bolt to the inner radius hole, and make the inner cut to complete the part.

Jig Tables

A MITER SAW or planer wouldn't be very effective or efficient if it lacked a sturdy surface to support the workpiece during cutting. In fact, a tablesaw wouldn't be worthy of its name without a table. But that doesn't mean you can't alter a machine tool's surface, or extend or augment its table in order to make the tool more useful.

For example, you can sand beveled edges on an oscillating spindle sander that lacks a tilting table by clamping a shopmade table atop the machine's stock table. You may choose to build a fancy table that adjusts to different angles, or you may just make a simple fixed-angle plywood table (see the bottom photo at right). In fact, performing angled operations with a shopmade tilting table is often easier than adjusting and readjusting the stock tilting table on a machine like a drill press. Tilt tables can also expand the range of work you can do on a machine. For example, a tilted-table jig used with a jointer allows you to cut clean raised panels without a shaper.

Another type of shopmade table that's extremely useful is an extension table. By effectively increasing the dimensions of a machine's table, an extension table offers better support for large workpieces. For example, attaching a shopmade extension table to a powered miter saw makes for easier and safer cutting, especially of long stock. And by incorporating a T-track flip stop system into

the tables, you get accurate, repeatable cuts as well.

You can also expand the versatility of portable power tools by using them with shopmade tables and jig setups. While jigsaws, belt sanders, and a variety of other tools make good candidates for table-mounted use, router tables are clearly the most popular applications, and you'll find several designs for them in this section

Tilt Tables

- Fixed-Angle Table (p. 122)
- Adjustable Tilt Table (p. 123)
- Angled Router Table (p. 125)
- Jointer Table (p. 126)

Extension Tables

- Miter Saw Extension Tables (p. 127)
- Bandsaw Extension Table (p. 128)
- Planer Extension Table (p. 129)
- Air Table (p. 130)

Power Tool Tables

- Router Table (p. 131)
- Router Biscuit Joinery Table (p. 133)
- Quickie Router Table (p. 135)
- Horizontal Router Table (p. 136)
- Belt Sander Table (p. 138)

If a machine's table isn't designed to tilt, a shopmade angled table lets you tilt the workpiece instead.

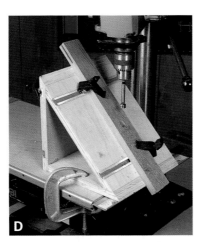

Fixed-Angle Table

This fixed-angle table will clamp to any drill-press table to hold parts at precisely 45 degrees—the most commonly drilled angle besides 90 degrees. In addition, it will hold parts vertically for end-boring operations.

In profile, the table is shaped like a 45-45-90-degree triangle. It's easy to make by simply cutting the parts from ¾-in.-thick plywood or MDF, sizing them to best suit your needs. (The table shown here is 9 in. wide and 9 in. high.) Glue the parts together as shown **(A)**, taking care to keep them precisely aligned while you drive nails or screws into the joints.

To facilitate securing workpieces to the table, add a pair of aluminum T-tracks that accept adjustable work clamps.

▶ See *"Extruded T-Tracks"* on p. 29.

Saw or rout a pair of grooves into the angled face, sizing them to accommodate your chosen T-track **(B)**. To maximize versatility when you're clamping workpieces of different sizes, the grooves should run side to side as shown. If the T-tracks have mounting holes, screw them into the grooves. If not, use epoxy to glue them in place, clamping them firmly at each end, and spanning each with a tape-covered stick (to resist glue) to distribute clamping pressure along the track **(C)**.

To use the table, attach it to your drill-press table with clamps or with bolts running through holes or slots cut into the base. Secure the workpiece to the angled table with a T-track clamp mounted in each track **(D)**.

[TIP] Filling in the irregular underside of a drill-press table casting with plywood makes it easier to clamp jigs and fixtures to the tabletop.

Adjustable Tilt Table

If you find yourself drilling holes at odd angles or needing to bore at compound angles, an adjustable tilting table is worth making. Designed to fit atop a stock drill-press table, this tilting table adjusts from horizontal up to 45 degrees of tilt. As shown in the drawing at right, the jig consists of a base that screws or clamps to the regular drill-press table, and a tilt table that's hinged to the base with regular butt hinges. A pair of adjustable plywood compasses attached to the sides of the table provide for solid support at your chosen angle. The jig base and table—made from ¾-in.-thick plywood or MDF—are sized so that the compasses will clear the edge of the drill-press table (A).

The jig's top includes a cutout for a replaceable table insert to minimize exit tearout on the workpiece. Use a jigsaw to cut out a 2¼-in.-sq. hole in the center of the tilt top, and then rout a ¼-in.-deep, ⅜-in.-wide rabbet around the hole to create the support ledge for the hardboard insert plate (B).

Using a bandsaw or jigsaw, cut the two compasses from ½-in.-thick plywood. Each is shaped like a 50-degree pie wedge with a radiused end. Rout a curved slot through each compass to allow table adjustment. Rout the radiused slots using a circle jig attached to a router or laminate trimmer, securing the compass atop a scrap of

(Text continued on p. 124.)

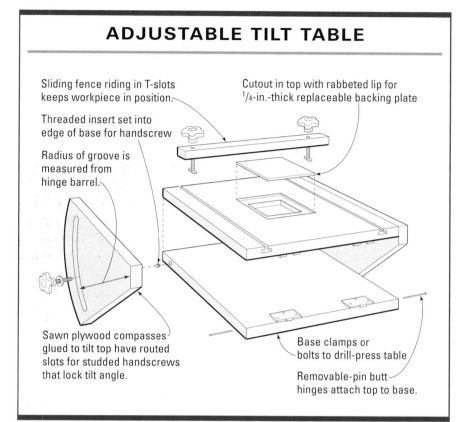

ADJUSTABLE TILT TABLE

Sliding fence riding in T-slots keeps workpiece in position.

Threaded insert set into edge of base for handscrew

Radius of groove is measured from hinge barrel.

Cutout in top with rabbeted lip for ¼-in.-thick replaceable backing plate

Sawn plywood compasses glued to tilt top have routed slots for studded handscrews that lock tilt angle.

Base clamps or bolts to drill-press table

Removable-pin butt hinges attach top to base.

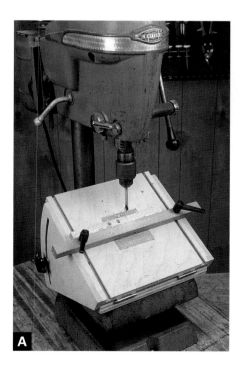
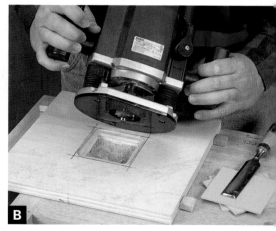

plywood with cleats or stop blocks **(C)**. (Here, the circle jig's pivot point engages a hole in the scrap piece that's clamped over the narrow end of the compass.) Before gluing and screwing the compasses to the sides of the tilt table, rout a pair of T-slots near the edges of the top to accept an adjustable fence. Then attach the tilt table to the front edge of the base with a pair of small butt hinges **(D)**.

► See *"Routing T-Tracks"* on p. 39.

After marking the location of the compass slot on the edge of the base, drill a hole and drive a threaded insert into each side to accept the threaded handscrews **(E)**. Finally, make a narrow fence from ¾-in.-thick stock, and drill a pair of holes through it, spacing them to meet the slots in the top. Fit T-bolts and handles through the holes to attach the fence.

Secure the tilt table to your drill-press table, and position the fence to support the workpiece so it doesn't slip down during angled-drilling operations. By angling the drill-press table as well as the tilt table, you can easily drill holes at compound angles.

[TIP] For efficiency and convenience, make a half-dozen or so insert plates at a time so you can replace them when they get ragged.

Angled Router Table

Here's a simple auxiliary table that tilts the workpiece 45 degrees relative to your router-table top. It's handy for angled routing, such as when you need to create an angled slot, shape a raised panel edge, or cut a finger groove for a drawer pull using a bullnose router bit **(A)**.

Start by cutting a piece of ½-in.-thick plywood or MDF that's as long as your router table's top and about 12 in. wide. Gauge a line 1 in. in from one edge, and then bore two overlapping 1-in.- or 1½-in.- dia. holes near the line's midpoint, spacing them approximately as shown **(B)**. Now rip the piece along the line to separate a 1-in.-wide strip, which will serve as the ledge to support the workpiece. Cut a shallow sawdust clearance groove in the table just where the ledge meets it, and glue and nail the two strips together, matching the position of the holes **(C)**.

For supports, saw a pair of right triangles from ½-in.-thick stock, with the hypotenuse cut to the angle of desired tilt. Also cut a clamping strip 2½ in. wide and a few inches shorter than the length of your tilt table. Working on a flat bench-top or machine table, carefully glue and nail the triangles to the ends of the clamp strip, and then attach the tilt table **(D)**.

Secure the tilt table to your router table fence with a pair of C-clamps, positioning the hole in the top at the location of the router bit. This way, you can reposition or lock down the tilt table using your router fence's existing hardware.

[TIP] **When boring with large-diameter Forstner® bits, slowing down the RPM of your drill helps the bit run cooler and cut more cleanly.**

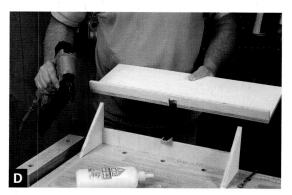

Jointer Table

This angled table jig allows you to make raised-panel cuts on the jointer by supporting the panel at an angle as it is fed past the jointer's cutterhead. Made from a piece of ½-in.- or ¾-in.-thick plywood or MDF, this table is supported at 15 degrees to the jointer tables by a couple of brackets that attach to the jointer.

You'll need to suit the table to your particular jointer, but that's not difficult. Begin by making the table large enough to support the largest panel you plan to cut. In order for the table to fit closely against the jointer tables, you'll need to notch the edge to clear the jointer's rabbeting ledge. Mark for the notch while holding the table piece against the rabbeting ledge (A), and then trim the piece as necessary. Next, design and make the support brackets (B). The jointer end of each bracket should be long enough and wide enough to screw or clamp to the jointer's base, cabinet, or to the machine itself.

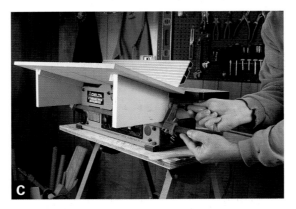

Once the supports are cut to shape, attach them to your jointer and fit the table in place (C). With the jointer set for a cut as deep as the desired height of the panel's central raised field (typically ⅛ in.), tap the table until its edge is precisely aligned with your jointer's infeed table. Then tighten the clamps or screws, glue and screw the table to the supports, and add glue blocks to reinforce the joints. Set the jointer's fence to the desired width of the beveled portion of the panel, and take several test cuts on scrap before cutting on your project (D).

> ⚠️ **WARNING** Always use push blocks—never your hand—to hold stock down as it passes over a cutterhead or blade.

Miter Saw Extension Tables

Miter saw extension tables provide side support for longer stock, and incorporating a fence and flip stops allows accurate, repeatable cutting. Begin by cutting the base platform from ¾-in.-thick plywood. Make it big enough to support the base of the saw and long enough to hold 18-in.- to 24-in.-long extension tables on each side. Screw the base atop a pair of 2x4 rails **(A)**, fastening additional 2x4 struts to the rails, if necessary, to accommodate the shape of your saw's base.

The extension tables are made from ¾-in.-thick plywood, each sized to match the width of the saw's table and high enough to sit flush with the saw table when mounted to the base wings. Each extension is constructed like a tunnel, with strips supporting a top piece **(B)**. The two rear strips that form the fence of each extension table are cut to match the height of the saw's fence and/or the length of the flip stops you wish to use with the saw. Doubled-up strips provide a wide edge for mounting T-track.

Screw the extension tables to the base. Bore a couple of pocket holes into the front and back pieces, but before screwing the tables to the base, check to make sure the tables are flush with the saw's table **(C)**. If necessary, add shims under the tables or saw base.

Mount T-track to each extension fence, and apply a self-stick measuring tape that's zeroed to the saw's blade. Even an 18-in.-long extension should allow you to set a flip stop to cut work up to about 30 in. long **(D)**. To cut shorter workpieces using the stop rail, employ an offset flip stop.

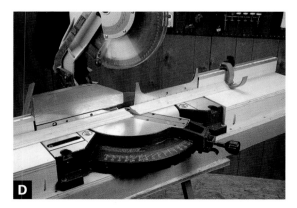

➤ See "*Variation*" on p. 193.

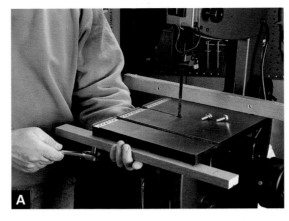

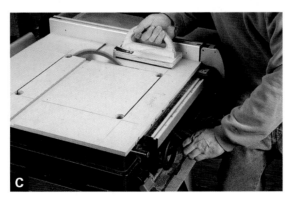

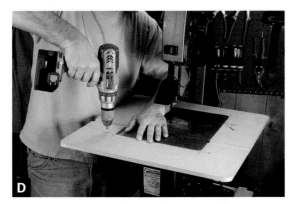

Bandsaw Extension Table

This bandsaw extension table is basically an MDF panel with a cutout in the center to accommodate your bandsaw's stock table. The panel is supported underneath by solid wood struts that extend beyond the saw's stock table. Size is a matter of preference.

Begin by making and attaching the struts. If your table already has threaded holes, you can use them to attach the struts **(A)**. If not, drill and tap two holes for each strut or clamp them to the saw itself. Next, cut your table panel from ¾-in.-thick MDF, plywood, or melamine. After carefully measuring your saw's table, mark the cutout to accommodate the table, while clearing the frame or any other obstructions.

For square or rectangular cutouts, drill a hole in each corner **(B)** that matches the radius of your saw table's corners and cut out the opening by plunge-cutting with the tablesaw or a portable circular saw **(C)**. When doing this on the tablesaw, hold the stock down firmly with a push block as you raise the blade into the stock. To prevent binding and kickback, don't cut the piece entirely free on the saw. Instead, stop the cuts short of the corners, and finish up with a jigsaw or handsaw.

Clean up the corners with a sanding block or rasp. If your bandsaw table has a miter gauge slot, cut a dado into the panel in line with the slot, oversizing it so it won't obstruct the gauge. Saw your way into the cutout area and slip the new table on the saw, trimming it as necessary for a snug fit. If necessary, place a shim atop the struts to bring the panel flush with the saw table. Finally, screw the panel to the struts **(D)**.

Planer Extension Table

This extension table simply sits atop your planer's existing table, adding support to make planing long boards easier and to help prevent sniping. To make the table, start by measuring the exact length and width of your planer's table, including its built-in extension wings, if any **(A)**. Then make the ¾-in.-thick MDF or melamine table wide enough to slide through the planer without hanging up. Its overall length isn't critical, but it should extend a couple of feet beyond your planer's existing tables and wings.

Now make the aprons that stiffen the overhanging sections. To calculate the length of the four side aprons, first subtract the exact total length of your planer's stock table and extension wings, and then divide that figure in half. Cut four 2-in.-wide pieces of solid stock to that length, and glue and nail or screw them to the underside of your new table at the sides of each end **(B)**. Then cut two more apron pieces to extend across the undersides of the ends, fitting them between the other aprons. In addition to stiffening the panel, the aprons engage the ends of your planer's table to keep the extension table in place **(C)**.

A neat addition to your new planer extension table is a diagonal fence bar. This is nothing more than a straight-grained stick with three dowels fitted along its length (one near each end and one in the middle). The dowels fit into holes bored into the extension table to angle the fence, as shown **(D)**. This fence is handy when you're planing narrow stock, since angling it uses a bit more of the width of the cutterhead. It also creates more of a shear cut, reducing tearout slightly.

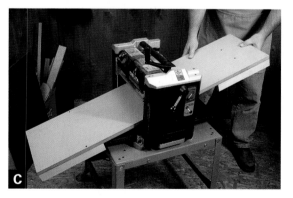

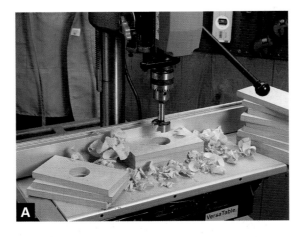

Air Table

An air table is a big, thin, perforated box that releases a cushion of air to allow heavy panels to glide over it with little effort. It's useful as an out-feed table for a tablesaw or other machine. The table is built like a torsion box. The one shown here is 3½ in. x 24 in. x 72 in., but you can make it any size you wish. After sawing the long sides, the central support ribs, and the inner strips, bore holes through all the inner strips and ribs **(A)** to allow the air to circulate inside the box. The top and bottom skins are cut from ¼-in.-thick single-sided melamine panel, with the melamine surface oriented to the outside of the box. The air holes in the top are drilled in a 2½-in.- x 2½-in.-grid pattern with a ⅛-in.-dia. drill bit **(B)**. Spacing isn't critical; just plan the grid layout so that holes aren't blocked by the inner structure.

As for any torsion box, assembly is best done on a large, flat surface. For long tables like this one, glue and nail components in sub-assemblies **(C)**, then put the sub-assemblies together, applying the top and bottom skins last.

Air is supplied to the table from a 3 hp–5 hp shop vacuum's blower port, which is connected to a plastic hose fitting mounted over a hole cut into the bottom center of the air table. Mount the table atop a pair of sawhorses or adjustable stands set next to your saw, aligning the table flush and level with the saw table **(D)**.

[TIP] Your shop vacuum can do double duty when you're using the air table with a benchtop-style tablesaw: Connect the blower hose to the air table and the vacuum hose to the saw.

Router Table

Few power tools in a modern woodshop are as useful or versatile as the router table. A basic table consists of only a few components: A top, a fence, a base, and an insert plate that is recessed into the top, to which the router is mounted. You can make these components yourself, or buy them ready-made. One option is to simply build the base, and then purchase a router tabletop or insert plate that includes features that would be impractical to build yourself, such as a built-in height-adjusting mechanism **(A)**.

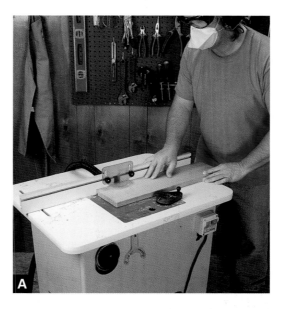

Many woodworkers prefer their router tables to be about as tall as their workbenches, but another strategy is to make it just shorter than the height of your tablesaw. That way, you can park the router table behind your saw and make it do double duty as an outfeed table when you're sawing large stock. The width and length of the base should be sized to allow the tabletop to overhang 3 or 4 inches all the way around.

Once you've made a dimensioned sketch of your desired table, cut the four sides of the base from ¾-in.-thick stock. I prefer MDF here, because its weight and density can help cut down the vibration and noise of the router. You can make each side from a single piece, but you'll use less stock if you glue up each side from three parts: an upper panel and two lower "leg" pieces **(B)**. A spline or biscuit joint between the legs and the upper panel adds plenty of strength to the joints. After all four sides are completed, glue and screw or nail them together to make the base. Glue and screw a triangular gusset near the bottom of each leg to make the feet. Drill a hole in the center of each triangle, drive in a T-nut, and then screw in an elevator bolt or threaded foot **(C)**. This allows you to level the router table and keep it from wobbling on an uneven floor.

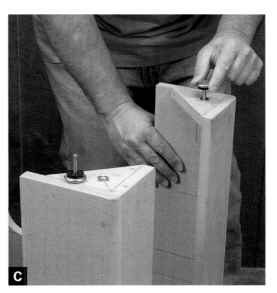

(Text continued on p. 132.)

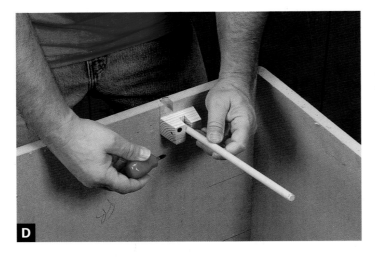

D

E

Hinging the router tabletop to the base makes it much easier to adjust and change router bits. First, make a support rod for the top from a sturdy dowel and screw it to a block of wood mounted inside the front edge of the table (D). To hinge the top, use butt hinges with removable pins. Set the top on the base, mark the hinge positions carefully, and then use a self-centering bit and drill to bore the holes, using the screw holes in the hinges as a guide. Remove the pins and screw the hinge sides to both the top and base (E). Set the top back in place and drive the pins home, and then mark and drill a shallow hole on the underside of the top for the dowel support rod.

To make your table safer and easier to use, purchase and mount a button-type on/off switch to the front of the base. Wire it to an electrical box fitted with a receptacle (to plug the router into) and mount it to the base (F). Attach your router to the insert plate, drop it into its recess in the top (G), plug it in, and you're ready to rout.

[TIP] You can use your router table with a jigsaw, portable spindle sander, or other power tool by making a custom insert plate that secures the tool underneath.

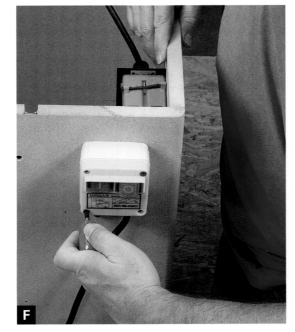

F

G

Router Biscuit Joinery Table

Plate joinery biscuits are great for joining parts, as long as they're fairly wide. But did you know you can also use round biscuits (Lamello® #11) to join parts as narrow as 2 in. The ⁵⁄₃₂-in.-wide slots for these are cut with a 2-in.-dia. slot-cutter bit used in a router table. This special router jig table mounts atop any router table. It has a pair of pivoting arms that swing the workpiece into the bit during slotting. It will work with rails ranging from 1¼ in. to 2½ in. wide; the eccentric end stop on the arms is reset to accommodate rails of different widths. One arm is used for right-hand frame stiles and the other for left-hand stiles **(A)**. Another fence mounted to the front of the jig is used to slot the end of frame rails.

Cut the jig table from a piece of ½-in.-thick plywood and drill the hole for the bit and the holes for mounting the arms, fence, and guard as shown in the drawing at right. Make the two pivot arms from straight-grained solid wood. For a smooth, precise pivoting action, mount the arms to the

(Text continued on p. 134.)

A

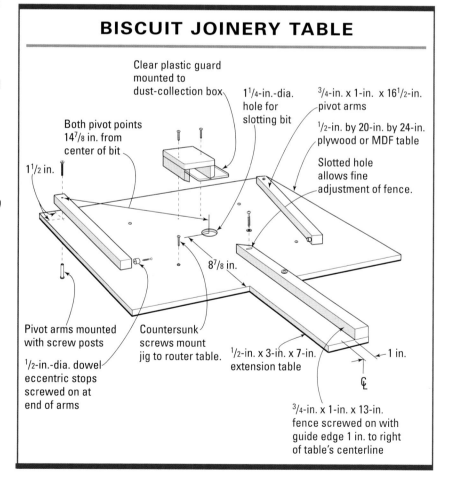

BISCUIT JOINERY TABLE

Clear plastic guard mounted to dust-collection box

Both pivot points 14⁷⁄₈ in. from center of bit

1½ in.

1¼-in.-dia. hole for slotting bit

¾-in. x 1-in. x 16½-in. pivot arms

½-in. by 20-in. by 24-in. plywood or MDF table

Slotted hole allows fine adjustment of fence.

8⁷⁄₈ in.

Pivot arms mounted with screw posts

½-in.-dia. dowel eccentric stops screwed on at end of arms

Countersunk screws mount jig to router table.

½-in. x 3-in. x 7-in. extension table

1 in.

₵

¾-in. x 1-in. x 13-in. fence screwed on with guide edge 1 in. to right of table's centerline

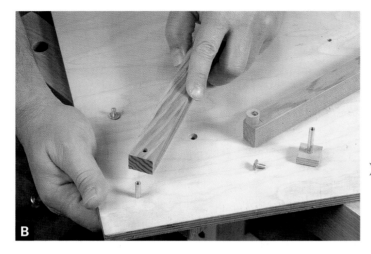

top using post screws **(B)**. The adjustable eccentric end stops are made from short lengths of dowel and mounted to the end of a pivot arm with a wood screw running through an off-center hole in each stop. This provides a bit of adjustment for fine-tuning the position of the slot on the ends of the stile workpieces.

▶ See *"Eccentric Stops"* on p. 195.

Glue a narrow ½-in.-thick plywood extension piece to the underside of the central fence to support the rail being cut. Screw the fence to the underside of the jig table **(C)**, slotting the innermost screw hole to allow fine adjustment of the fence for tweaking the slot location in the end of the rail.

A clear guard, made from a 3 in.-sq. piece of ⅛-in.-thick Lexan, protects hands from the cutter. It screws to the overhanging top of a three-sided plywood box that holds the guard about 1 in. above the table. The box doubles as a dust collection port.

For quick and precise mounting, the table screws to the top of your router table with screws driven near each corner of the jig **(D)**. Set the cutter height to slot the workpiece approximately halfway across the thickness of the stock. To ensure that the slots match up accurately, make sure to orient each workpiece face up on the jig.

Quickie Router Table

This router table uses a pair of quick-operating clamping cleats for speedy mounting and dis-mounting of a router. The table, which is sized for smaller tasks, clamps into a workbench vise so it isn't taking up shop space when not in use **(A)**.

The top of the router table is made from a 12-in.-square piece of ¼-in.-thick hardboard or tempered Masonite. To strengthen the thin top and align the router underneath, glue strips of ½-in.-thick plywood to the underside. Size the strips to create a frame around the base of the router, thus capturing and centering it **(B)**. (Biscuit-join the strips together for additional strength.) After boring a 1-in.-dia. bit opening through the center of the hardboard top, saw one edge of the table to an 11½-in.-radius on the bandsaw **(C)**. (The radius is necessary to accommodate the swing of the table's pivoting fence.)

Install a pair of threaded inserts into two of the strips to hold the cleats. Center the inserts and space them ½ in. or so from the inside edge of the frame. Cut two 1½-in. x 3-in. cleats from ½-in.-plywood and drill them out for the handscrew studs. Cut one edge of each cleat concave to facilitate router removal **(D)**. To enhance clamping power, cover the underside of the cleats with anti-slip tread tape.

The top attaches to a vertical support made from ¾-in.-thick plywood cut to 12 in. wide and 13 in. to 16 in. long. Glue and screw the table to the top edge of the support, adding a pair of 8-in. triangular plywood gussets for reinforcement **(E)**.

The pivoting fence is a 16-in.-long piece of 2-in. x ¾-in. solid wood. Drill a countersunk hole near one end to accept the pivot screw, and install a hanger bolt for a locking hand knob at the other end **(F)**. Cut out an arched bit-clearance hole on the bottom edge at the table hole location. Using scrap from the earlier bandsaw radius cut, make a 1¾-in.-long spacer, and glue it to the underside of the fence bar **(G)**. Install a fender washer between the spacer and the knob to complete the clamping mechanism.

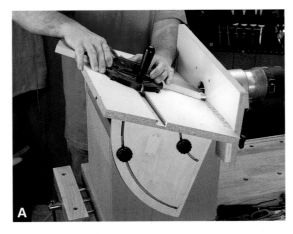

Horizontal Router Table

A router table that orients the router horizontally instead of vertically is terrific for raising panels, cutting joinery, and other shaping jobs where you'd rather have the work flat on a table than standing vertically against a fence. This horizontal table has a twist: It can be tilted, allowing a greater variety of profile cuts, panel-shaping options, and configurations for cutting angled joinery **(A)**.

Begin by making the table base from ¾-in.-thick MDF or particleboard. Cut out three 12-in.-sq. pieces for the front and sides, and a 12-in. by 3-in. strip for the bottom rear piece. Using a bandsaw or jigsaw, cut a 3-in.-high, ¾-in.-deep notch at the top back corner of each side, to accept the cross member. Glue and nail the base pieces together, attaching them to a 12-in. by 18-in. bottom piece **(B)**. Make sure the base is square and that all the top edges are flush.

Use ¾-in.-thick plywood for the 3-in. by 18-in. cross member, and drill holes each ¾ in. down from the top edge and ¾ in. in from the end of the member, for two ¼-in. hangar bolts. Install the bolts **(C)** and then glue and screw the cross member into the notches on the base.

Make the 9-in. by 18-in. router plate from two pieces: a ¼-in.-thick melamine or tempered hardboard face, and a ½-in.-thick MDF back. Using the router's sub-base as a template, drill mounting holes in the face piece to attach the router. Then jigsaw a hole in the back piece large enough to clear the router's base **(D)**. Carefully align and glue the two halves together.

Drill one end of the plate for the pivot screw hole and then rout the radiused slot at the opposite end. Locate the pivot screw hole 3 in. from the bottom and ¾ in. from the side edge of the plate.

The ¼-in.-wide compass slot is routed with a 16½-in. radius centered on the pivot hole. Mount the plate with a pair of threaded handscrews **(E)**. Loosening the handscrew on the plate lets you adjust the distance of the router bit relative to the table surface.

Lay out the two compasses, which support the tabletop and allow it to tilt. Both are cut from a single ½-in.-thick plywood blank. Temporarily screw the blank to scrap plywood, and attach scrap blocks to provide pivot points for a router trammel **(F)**. Rout all four slots at the radiuses shown in the drawing at right, and then cut the compasses from the blank.

Next, cut the 14-in. by 22-in. tilting tabletop from ¾-in.-thick melamine or MDF, beveling the underside of its rear edge at a 45-degree angle, for clearance when tilted. Plow two ½-in.-wide, ⅜-in.-deep dadoes into the underside of the top **(G)**. Space them exactly 13½ in. apart, so that the inside faces of the compasses will be flush against the sides of the base. If you plan to use a miter gauge with the table, plow a groove for the track now.

Glue the compasses into their dadoes, aligned as shown in the drawing at right. Finally, install four threaded inserts on the sides of the base to accept the studded handscrews that secure the top at an angle. Set the top-and-compass assembly on the base, bringing the top's beveled edge against the router plate. Mark the upper end of each compass slot and drill on those marks for the threaded inserts. Then install the studded handscrews **(H)**, and you're ready to rout.

[TIP] To keep parts aligned during assembly, drive in small brads and then clip their heads off nearly flush. The projecting nibs keep the slippery glued surfaces from sliding while clamps are tightened.

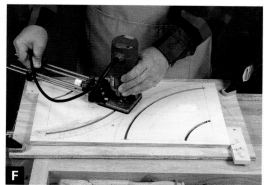

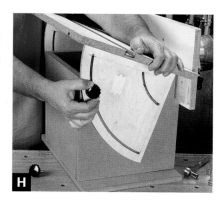

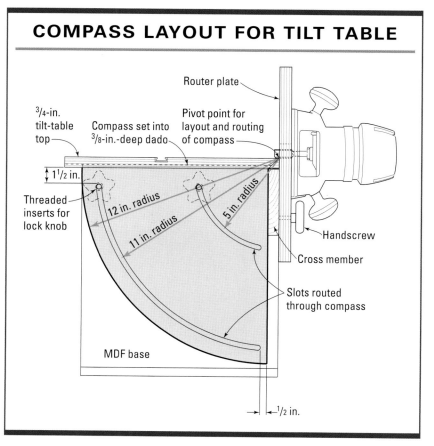

COMPASS LAYOUT FOR TILT TABLE

Router plate · ¾-in. tilt-table top · Compass set into ⅜-in.-deep dado · Pivot point for layout and routing of compass · Handscrew · Cross member · Slots routed through compass · 1½ in. · Threaded inserts for lock knob · 12 in. radius · 11 in. radius · 5 in. radius · MDF base · ½ in.

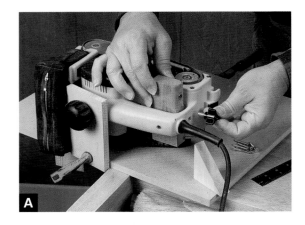

Belt Sander Table

This belt sander table consists of a regular belt sander clamped to a base with a table mounted in front of it. The table is angled to distribute sanding belt wear more evenly.

To make the table, begin by cutting a base plate out of ¾-in.-thick plywood or MDF that's about 15 in. wide and at least 8 in. longer than your sander. If your sander has a detachable front knob or mounting holes, use these to secure the unit to the base with mounting blocks. For best stability, locate blocks fore and aft of the sander (A). To make the sander easier to remove, use threaded inserts and handscrews or knobs, if possible. Cut and attach the mounting blocks to the sander first, and then glue and screw them to the base with the sander in place, checking that the sander's platen and belt surface are square to the base. If necessary, add spacers under the sander to stabilize it.

Make the angled work table about 8 in. wide and about 2 in. shorter than the length of the base. Cut two end strips to position the table slightly above the edge of the belt at the low end and 1½ in. below the top edge of the belt at the high end. Glue and nail or screw the ends to the table (B). Before screwing the table to the base from underneath, "cut in" the edge of the table by pressing it gently against the spinning belt (C). This ensures a minimal gap between the belt and table. Before using your belt sander table, clamp or screw the base to your benchtop (D).

> ⚠ **WARNING** When sanding, hold parts firmly down on the table and avoid orienting sharp corners against the direction of belt travel.

Jigs for Portable Tools

STATIONARY AND BENCHTOP machines aren't the only woodshop tools that benefit from the use of jigs and fixtures: You can also expand the repertoire of your portable power tools and hand tools by using them with a variety of jigs and fixtures. Take fences used with routers and circular saws, for example: These range from simple crosscutting guides used freehand to make quick, accurate square and miter cuts on construction lumber, to long panel fences and railed guides that clamp to large panels, allowing you to cut big sheets of plywood to manageable size. If you don't have a big panel saw at your disposal, you'll benefit from building the stationary crosscutting jig in this section. It converts a portable circular saw into a sort of radial-arm saw, enabling mitering and crosscutting of wide stock. Other guides include a circle jig for cutting curves with a jigsaw or router, and a jig for quick and easy biscuit-slotting of wide panels.

The router is clearly one of the most valuable tools in the woodshop, and there is an endless variety of router guides and jigs that make it even more versatile. One example is a router guide that facilitates everything from basic grooving to making tapered cuts for sliding dovetails. Others include a parallelogram fence jig that auto-

Fences and Guides

➤ Crosscut Guide (p. 141)
➤ Panel-Cutting Fences (p. 142)
➤ Railed Guide (p. 144)
➤ Crosscut Jig (p. 145)
➤ Circle-Cutting Jig (p. 146)
➤ Biscuit Joiner Guide (p. 148)

Router Jigs

➤ Basic Router Guide (p. 150)
➤ Even-Spacing Fence (p. 151)
➤ Joinery Jig (p. 152)
➤ Panel-Hollowing Jig (p. 154)
➤ End-Routing Jig (p. 156)
➤ Fluting Jig (p. 157)
➤ Pattern Routing Jig (p. 159)

Router Bases

➤ Four-Sided Base (p. 161)
➤ Self-Centering Base (p. 162)
➤ Angled Base (p. 163)
➤ Edge-Routing Base (p. 164)

Hand Tool Guides

➤ Shooting Boards (p. 165)
➤ Small Crosscut Box (p. 166)

▶ A ROUTER SUB-BASE JIG

Attaching your router to a new base or jig setup is a snap if you simply use its stock plastic sub-base as a template for boring the mounting holes for your jig. Remove the sub-base and apply a little adhesive-transfer tape (see p.173) or double-stick tape to the side that contacts the router. Press the sub-base onto

the underside of the new jig base or plate. Flip the pieces over and use the countersunk holes in the sub-base to drill holes through the new plate—an easy job using a self-centering bit chucked in a portable drill. If the new jig base or plate is thicker than the sub-base, you may have to counter-bore your holes or use longer mounting screws.

This versatile mortising jig uses a top-mounted template and a bushing-guided router to cut mortises in straight, angled, or curved stock.

Clamping a simple wooden guide strip to a handplane facilitates tasks such as edge jointing and camfering.

matically spaces bookcase shelf dadoes, and one that is used for hollowing or shaping concave panels. There's also a neat pattern-routing setup for creating signs and decorative designs, a fluting jig for enhancing turned spindles on the lathe, and a mortising jig (see bottom left photo) that can accommodate square, mitered, or curved workpieces.

An even simpler way to squeeze more performance out of any router is to replace its sub-base with a jig base or mounting plate. One such base lets you space the router at up to four different distances from a fence or guide, simply by rotating the tool. Another jig base automatically centers the bit across the thickness of the stock for making centered grooves or mortises. Yet another replacement router sub-base tilts the router on the work surface at either of two different angles.

While this section focuses primarily on portable power tools, it would be a mistake to overlook jigs that expand the functions of basic woodworking hand tools. Therefore, you'll also find plans for making a traditional shooting board used for trimming the ends of frame members and parts to a perfect 90 or 45 degrees. And, there's a special crosscut box that's sized and designed for safely and very accurately cutting small wood dowels and strips to exact length using a fine-toothed handsaw.

Crosscut Guide

Sometimes when cutting lumber for shop furniture or backyard construction projects, you just need a quick crosscutting guide. This one will help you accomplish both square and 45-degree miter cuts using a regular portable circular saw. For the jig's fences and crossbar, cut three 20-in. lengths from ¾-in.-thick, 2½-in.-wide solid wood stock. Plow a ¼-in.-deep, 2½-in.-wide dado on the underside of the square-cutting fence 6 in. from one end, and dado the bottom of the miter fence at 45 degrees in each direction **(A)**.

Miter one end of the crossbar so it comes to a point (like a fence picket). One end of both the miter fence and pointed crossbar is drilled and joined with a carriage bolt and hand knob so that the fence will pivot 45 degrees in either direction **(B)**. Glue the other end of the crossbar to the dado in the square-cutting fence. With a jigsaw or bandsaw, cut a handle for the jig from a 9-in. x 4½-in. piece of ¾-in.-thick plywood **(C)**. The twin handle extends in each direction to allow both right- and left-handed cuts. Attach the handle to the center of the crossbar with screws from underneath. To use the jig, pull the crossbar tight against either edge of the stock with one hand, and use the appropriate fence to guide the saw through the cut **(D)**. For 45-degree cuts, pivot the miter fence as necessary.

[TIP] Rub a little wax onto the edge of any wood fence or guide to help tools glide more smoothly along the edge.

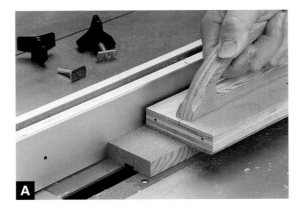

Panel-Cutting Fences

It makes sense to cut large, heavy MDF and ply-wood sheet goods down to manageable sizes before trying to hoist them onto the tablesaw for further cutting. This panel-cutting fence, which is long enough to crosscut standard 4-ft. x 8-ft. sheets, features quick-action clamps and a crossbar that aligns the fence for square crosscuts. Make the 2¼-in.-wide, 56-in.-long fence bar from straight-grained solid wood. The first step in making the T-slot in the fence (for the stop block) is to cut a ½-in.-wide, ⁵⁄₃₂-in.-deep centered groove along the entire length of the strip **(A)**. Then cut a 7-in. x 56-in. baseplate from ¼-in.-thick hardboard, and glue and nail it to the grooved side of the fence **(B)**. Next, complete the T-slot by cutting a ¼-in.-wide slot through the base, centered on the previously cut groove **(C)**.

Make the clamping stop block from a 2½-in. x 6-in. piece of ¾-in.-thick plywood **(D)**. Cut a ¼-in.-wide, ⅜-in.-deep dado across the width of the stop, then glue a ¼-in. x ¾-in. x 1½-in.-long wood spline into one end of the dado. (The spline will keep the stop aligned in the T-slot.) Drill a hole through the stop, centered on the spline-less section of the dado, and install a T-nut and threaded handscrew (or a T-bolt and hand knob.) Attach a small quick-action clamp atop the block opposite the handscrew/hand knob, and slide the stop into the T-slot.

Using ¾-in.-thick plywood, make the 3½-in. x 24-in. crossbar for the other end of the fence. If you'll only use the jig for square cuts, simply screw the crossbar to the underside of the fence bar, making sure it's positioned exactly square. However, if you'd like to occasionally remove the bar to make angled cuts, then fit it with a spline and T-nut or T-bolt as you did the stop block. In either case, screw another small quick-action clamp to the crossbar, centered on the T-slot **(E)**.

Clamp the fence to your workbench and "cut in" the hardboard baseplate, trimming it exactly flush with the blade teeth **(F)**. This allows you to position the jig on the workpiece simply by aligning the baseplate with your cutline. To use the fence, position it atop the work with the crossbar pressed firmly against the edge of the panel. Slide the stop block tight against the opposite edge of the work, tighten the handscrew, and engage the quick clamps at both ends. Make sure to keep the saw's base firmly against the fence as you cut **(G)**.

Even though the baseplate is trimmed for use with a particular circular saw, you can still use the opposite edge of the fence to guide a router or jigsaw. To facilitate setup, make a spacer strip from thin hardboard **(H)**. Its width should equal the distance from the edge of the power tool's base to the near edge of the cutter or teeth.

[TIP] To keep the slide-on crossbar perfectly square to the fence, drive a screw through the baseplate into the crossbar.

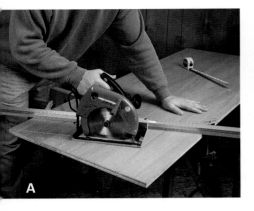

Railed Guide

An alternative to the panel fence shown on the previous page, this cutting guide **(A)** uses a rail that engages a sliding base mounted to your circular saw. Although it takes more effort to make and mount, this special baseplate prevents the saw from drifting off course, as is possible with a standard fence guide. The rail for this guide is a commonly available 6-ft.-long aluminum sliding door track with a pair of ½-in.-wide channels.

Begin by screwing the track atop a 2½-in.-wide, 6-ft.-long, ½-in.-thick plywood clamping strip **(B)**. Next, cut the sliding saw base from ¼-in.-thick hardboard, making it as long as your saw's sole and ¾ in. wider. On one long edge of the base, glue and nail a ¾-in. x ¹⁵⁄₁₆-in. riser strip, topped by a 1½-in.-wide, ¼-in.-thick hardboard guide bar mount **(C)**. Attach the base to the saw with four short flat-head screws through countersunk holes in the hardboard base. Make the 12-in. x ⅝-in. x ½-in. guide bar from UHMW plastic stock **(D)**.

▶ See *"Slippery Plastics"* on p. 20.

Taking very light passes, joint or plane the bar until it fits snugly in the track without binding. Bevel its leading edges slightly, so they don't hang up. Finally, position the saw baseplate next to the rail with the blade exactly parallel to the track, and then screw the base to the guide bar **(E)**.

To use the jig, clamp the track to the panel, making sure it extends at least a couple of inches at both ends. Then slip the guide bar in the track and cut.

> ⚠ **WARNING The baseplate may defeat the saw's blade guard, so use extreme caution when operating the saw.**

Crosscut Jig

This jig allows you to crosscut and miter panels up to ¾ in. thick and 24 in. wide using a portable circular saw **(A)**. You'll size this guide rail assembly to accommodate the width of the saw's baseplate.

Begin by cutting a pair of 45-in.-long, 1½-in.-square pieces for the side rails. Then make a 1½-in.-thick x 3-in. pivot block that's exactly as long as your saw base is wide. Also make an end piece that's the same length, but 1½-in. square. Join these parts atop a 45-in.-long strip of ¼-in.-thick hardboard, gluing and nailing or screwing the entire assembly together **(B)**. Also cut a ¾-in. x 3½-in. support block that's as long as the width of the rail assembly, and glue it to the underside at the far end.

Clamp the completed assembly to sawhorses, and "cut in" the base plate with your circular saw. Drill a ⁵⁄₁₆-in.-dia. hole through the pivot block, in line with the saw kerf and centered across the width of the block. Make a 48-in.-sq. baseboard for the jig from ¾-in.-thick MDF or plywood, and mark a pivot point 16 in. from the baseboard's left side and 1½-in. from its back edge. From this point, use a circle jig (see "Circle-Cutting Jig" on p. 146) set to a 42-in. radius to rout a ¼-in.-wide arcing slot across most of the panel. Glue and nail a 3-in.-wide fence 6 in. from and parallel to the back edge of the baseboard, and a 6-in. x 6-in. riser block centered over the pivot hole. Drill through the block and baseboard at the pivot hole, and then attach the guide rail with a carriage bolt **(C)**. Use another carriage bolt and a hand knob to lock the angle of the rail at the opposite end **(D)**.

[TIP] **For frequently used angles, add stop blocks or mark angles on the jig's base.**

CIRCLE CUTTING JIG

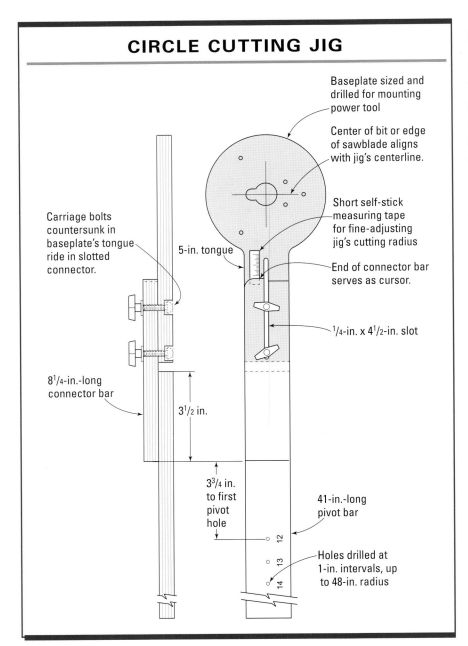

Baseplate sized and drilled for mounting power tool

Center of bit or edge of sawblade aligns with jig's centerline.

Short self-stick measuring tape for fine-adjusting jig's cutting radius

End of connector bar serves as cursor.

Carriage bolts countersunk in baseplate's tongue ride in slotted connector.

5-in. tongue

$^1/_4$-in. x $4^1/_2$-in. slot

$8^1/_4$-in.-long connector bar

$3^1/_2$ in.

$3^3/_4$ in. to first pivot hole

41-in.-long pivot bar

12

13

14

Holes drilled at 1-in. intervals, up to 48-in. radius

Circle-Cutting Jig

Cutting a perfectly circular disc of wood isn't a particularly daunting challenge—as long as you use a good circle-cutting jig. This one is useful for sawing or routing arcs and rounds up to a whopping 4-ft. radius. As shown in the drawing at left, it consists of three parts: A banjo-shaped baseplate that mounts to the power tool, a long bar with various pivot holes for swinging the tool, and a connector that joins the parts, creating a sliding joint that allows fine adjustment of the cutting radius.

Cut the baseplate/compass bar from a 6-in. x 11-in. piece of $\frac{1}{2}$-in.-thick plywood. First, make two parallel cuts on the tablesaw to form a 5-in.-long, 2-in.-wide compass bar on one end of the plywood **(A)**. Then use the bandsaw to cut out the round section of the baseplate **(B)**. On the drill press, bore a pair of countersunk holes in the bar for two short $\frac{1}{4}$-in. carriage bolts. Also bore mounting holes for attaching a jigsaw, as well as a clearance hole for the blade. For proper cutting action, the jigsaw should be positioned to align the sawteeth with the baseplate centerline **(C)**. If you plan to use a router with the jig, drill mounting holes for it now too.

From a 2-in.-wide strip of $\frac{1}{2}$-in.-thick plywood, cut a 41-in.-long piece for the pivot bar, and an 8¼-in.-long piece for the connector. Using a dado blade in the tablesaw, cut a ¼-in.-wide, 4½-in.-long slot centered in the connector. Cut the plywood on one side of the slot back ¼ in., and mark the tip of the other with black marker pen (the tips act as a pointer for

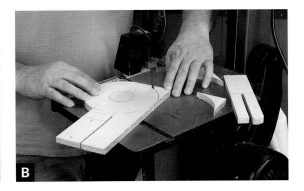

fine adjustments). On the pivot bar, mark out 37 pivot holes and drill them with a ⅛-in.-dia. bit chucked in the drill press **(D)**. Glue and nail the connector to the end of the pivot bar, overlapping it 3½ in., as shown in the drawing. Apply a short length of self-stick measuring tape to the top of the base plate's tongue **(E)**. (You can use a scrap of measuring tape left over from another jig project, as you'll use only the fractional graduations on the tape for making fine adjustments.) Finally, slide the slotted end of the pivot bar onto the carriage bolts on the base and secure the connection with a pair of hand knobs.

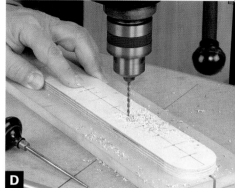

To use the jig, choose the pivot hole that works best for the desired radius of the cut, and measure from it to the sawblade or router bit. To adjust the exact radius, slide the pivot bar in or out as necessary. Starting with the blade at the edge of the work, pivot the jig around smoothly to make the cut **(F)**. It's best to take a trial cut and check the radius; if it's off, you can use the self-stick scale and cursor to tweak the setting by measured increments. Used with a router, the circle jig is very useful for cutting radius slots used in many pivoting jigs.

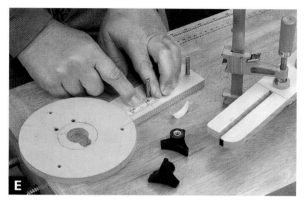

▶ See *"Adjustable Tilt Table"* on p. 23.

It can also be used to create curved recesses for line inlays, or for decorative shaping **(G)**.

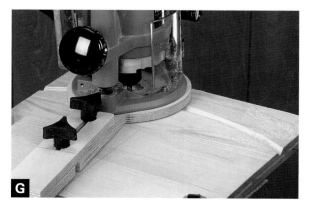

BISCUIT JOINER GUIDE

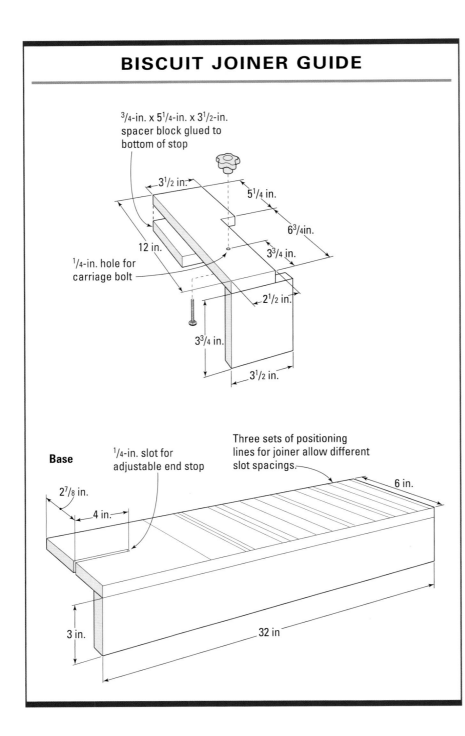

³/₄-in. x 5¹/₄-in. x 3¹/₂-in. spacer block glued to bottom of stop

3¹/₂ in.

5¹/₄ in.

6³/₄in.

12 in.

3³/₄ in.

¹/₄-in. hole for carriage bolt

2¹/₂ in.

3³/₄ in.

3¹/₂ in.

Base

¹/₄-in. slot for adjustable end stop

Three sets of positioning lines for joiner allow different slot spacings.

6 in.

2⁷/₈ in.

4 in.

3 in.

32 in

Biscuit Joiner Guide

Marking the location of biscuit-joinery slots on multiple cabinet parts is a less tedious affair when you use this jig to position and space slots evenly on the edges and faces of parts, thus eliminating the need to mark individual parts. A sliding end stop allows you to regulate the distance between the first slot and the edge of the work.

Assemble the stop from pieces of ½-in. and ¾-in.-thick plywood, as shown in the drawing at left. The guide itself is made from two 32-in. lengths of ¾-in.-thick plywood: a 6-in.-wide base, and a 3-in.-wide face strip. Cut a 4-in.-long slot for the sliding end stop, centering it in the left-hand end of the base **(A)**. A carriage bolt and hand knob secure the end stop in this slot. Glue and nail the base to the edge of the face strip. Draw a line ¾ in. from the back edge of the base along its entire length. Then draw a black line across the base 9 in. from its slotted end. The right-hand edge of your biscuit joiner aligns with this line when you'll cut the first biscuit slot on the left-hand side of the workpiece. Continue marking guidelines across the base at 3½-in. intervals from the first line. For versatility in spacing biscuit slots, mark additional guidelines at 4-in. and 4½-in. intervals **(B)**, using other colors.

A

For edge-slotting two parts—say, to join them end-to-end or side-to-side—position the end stop so that the first and last slots will be spaced as desired relative to the ends of the workpiece **(C)**. Lay the workpiece on a ¾-in.-thick scrap of plywood or particleboard, and then clamp the edge of the workpiece to the base, aligning it to the long black line. Set your biscuit joiner with its right edge aligned with the first guideline, and plunge a slot. Continue plunging slots, positioning the machine relative to the guidelines of one color **(D)**. Always position the biscuit joiner's right edge along the guideline, although exact positioning isn't critical because the slots allow a bit of side-to-side play. Repeat the process on the mating workpiece, making sure to place the biscuit joiner base against the same face of the workpiece so the parts will align properly, even if slots aren't exactly centered across the thickness of the stock.

For cabinet parts that join at right angles, such as a shelf to a cabinet side, set up your slotting spacing and cut the edge slots as usual. Now, without changing the end-stop setting, flip the guide jig up so that its face strip sits atop the panel. Position the jig so the end stop lines up with the edge of the workpiece, and clamp it down. Then vertically plunge slots into the panel, using the same set of guidelines you used for edge slotting **(E)**.

B

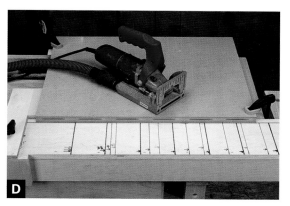

C

D

E

Basic Router Guide

This basic router guide is an extremely versatile jig, allowing you to rout rabbets and dadoes, stopped slots, and more on stock up to 24 in. wide. The guide's adjustable fence not only allows the use of routers with different-diameter bases, but also lets you rout oversized grooves and tapered slots for sliding dovetails.

Make each of the jig's twin fences by gluing a ¾-in. x 2-in. x 32-in. solid wood strip to a 4-in. x 32-in. strip of ¼-in.-thick hardboard. Next, make two 1-in. x 3-in. x 12-in. solid wood cross members, and saw a ⁵⁄₁₆-in.-wide, 5-in.-long centered slot on one end of each. You can saw the slot using a dado blade, or cut each slot in two passes with a regular blade **(A)**. Glue and screw the un-slotted ends of the cross members to each end of one of the fences, making sure they're perfectly square **(B)**. Bore a ⁵⁄₁₆-in.-dia. centered hole at each end of the other fence, and attach it to the cross members with carriage bolts and hand knobs fitted in the slots **(C)**.

Prepare the router guide for standard cuts by spacing the fences to capture the router's base snugly while allowing it to slide smoothly. Square the jig up to the work by butting one of the cross members against the workpiece edge. Clamp the jig in place and rout **(D)**. To make a dado that is larger than the diameter of your bit, simply reposition the adjustable fence as necessary, and make the cut in two passes instead of one. For tapering sliding dovetail slots, angle one of the fences slightly to produce the desired degree of taper **(E)**.

Even-Spacing Fence

Routing evenly spaced dadoes in a case to accept shelves or dividers is a breeze with this even-spacing fence. The jig is made up of only four wood strips, joined together to form an adjustable parallelogram. The jig shown here was sized to rout dadoes spaced up to 12 in. apart on panels up to 15 in. wide, but you can build one as large or small as you wish.

On the tablesaw, rip two 1½-in. x 20-in. strips from ¾-in.-thick straight-grained stock. Also cut the two 1-in. x 12-in. connecting struts from ½-in.-thick stock. On the drill press, bore a ¼-in.-dia. hole at both ends of each of the four pieces. Center the holes and space them 3 in. from the ends of the fences and ½ in. from the ends of the struts. Counter-bore the fence holes to recess the carriage bolt heads. On the underside of each strut end, apply a small piece of self-stick coarse-grit sandpaper to help hold the jig adjustment once it's set. Use a razor knife to cut out the paper around the holes **(A)**. Drive the 1½-in.-long carriage bolts into their fence holes and slip the struts in place, securing each with a pair of hand knobs **(B).** Finally, cut an alignment strip to match the size of the desired dado, and screw it to the bottom of one of the fences **(C)**.

Set the jig to the desired dado spacing, accounting for the width of the router's base and the bit diameter. Clamp the fence atop the workpiece, and rout the first dado. Next, reposition the jig, locating its alignment strip in the previously routed dado, and rout the next dado **(D)**. Repeat the process for all subsequent cuts.

A

B

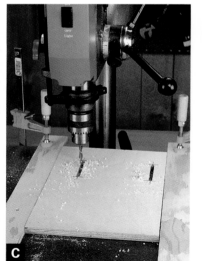

C

D

Joinery Jig

In addition to standard edge-mortising, you can use a plunge router to cut accurate mortises in the ends of frame members, for loose tenon joinery, or to cut half-lap joints, slots for splines, and even to shape tenons. This jig uses a plunge router fitted with a guide bushing that's guided by a hardboard template. A jig base supports and positions the template relative to the workpiece. To cut various joints, you make and attach different templates to the jig. The workpiece is aligned to the template with an adjustable fence, which also tilts for cutting angled joinery, as shown in the bottom left photo on p. 140. A quick-action clamp secures the work during routing.

Start by making an 8-in. x 12- in. template support from ½-in.-thick plywood. With a jigsaw or bandsaw, cut a 4-in. x 5-in. notch out of this support **(A)**. Also from ½-in.-thick plywood, make a 4-in. x 12-in. top for the jig. On the tablesaw, cut two ¼-in.-wide, 2½-in.-long slots into the top, spacing them approximately as shown **(B)**. Drill a pair of ¼-in.-dia. holes through the template support 2 in. from the unnotched edge and spaced to match the slots in the top. Counter-bore these holes from the top to recess the heads of two carriage bolts that attach the template support to the jig top.

Make the jig's 8-in. x 12-in. vertical work support from ½-in. or ¾-in.-thick plywood. Also make a pair of 3½-in. triangular gussets to brace the top. Outfit your drill press with a ¼-in.-dia. straight router bit (preferably spiral-fluted), and set your machine to its highest speed. Clamp a couple of guide boards to your drill-press table 12 in. apart, and rout a pair of 3-in.-long slots through the work support **(C)**. Start the slots about 2 in. from the left-hand edge, and space them from the top edge 1 in. and 6 in., respectively.

Glue and nail all the jig base parts together, taking care to keep the top perfectly square to the work support **(D)** by gluing the triangular gussets in place. Once the assembly dries, attach the template support to the top with short carriage bolts and small hand knobs. Now screw the hardboard template (made as described in Section Nine) to the top with four or five short screws **(E)**. Cut a 1½-in. x 12-in. workpiece fence from ¾-in.-thick stock, and miter its left-hand top corner. Rout a pair of ¼-in.-wide, 1-in.-long centered slots through the fence, spacing them to match the slots in the work support. (The mitered corner and mounting slots allow the fence to be tilted when necessary.) Mount the fence to the jig base with a pair of ¼-in. carriage bolts inserted through the slots and secured with hand knobs **(F)**. Finally, bolt a large quick-action clamp (screwed to a spacer block) to the right-hand side of the work support.

Clamp the workpiece in place, adjusting the fence so the work is positioned side-to-side as desired under the template. Then adjust the template support to reposition the template as necessary to make the desired cut. To prevent chip clogging, it's best to clamp the work so its top end is slightly below the template **(G)**. You can also rout joinery on curved or irregularly shaped pieces by making a fence that matches the shape of the workpiece **(H)**.

[TIP] To rout a mortise into the edge or face of a workpiece, use the jig without a fence.

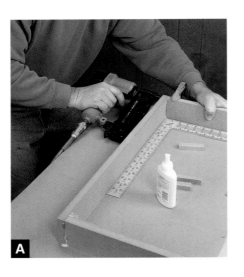

Panel-Hollowing Jig

A router used in conjunction with a panel-shaping jig can hollow planks and panels or create convex surfaces. The jig consists of a router track, a carrier frame, and a base to which the carrier frame and workpiece mount.

Build the carrier frame from ¾-in.-thick plywood or MDF. Its outer dimensions should be about 10 in. wider and 12 in. longer than the largest intended workpiece. (This frame is 24 in. x 34 in.) Make the frame members at least 2¾ in. wide (wider for very deep curves). Assemble the frame on a flat surface, gluing and nailing or screwing the butt-joined corners together and reinforcing them with glue blocks **(A)**.

Make the router track next. For its outer frame, use 1½-in.-wide, straight-grained, ¾-in.-thick stock. Make the inside width of the router track ¼ in. wider than the diameter of your router's baseplate. Also make a pair of ⅜-in. x ¾-in. hardwood rails that attach to the inside long edges of the side frame pieces to support a replacement router sub-base. Glue up the entire track, assembling it square and flat **(B)**.

Make the square sub-base from ¼-in.-thick hard-board, sizing it to fit snugly between the side track frame pieces without binding. Drill countersunk holes to mount the sub-base to your router, posi-tioning it so that it won't hang up on the frame **(C)**.

Saw out two sets of curved guides to support the ends of the router track. The upper half of each set mounts onto the ends of the router track, while the lower half—with its complementary curve—attaches to the carrier frame **(D)**. The radius of the curves reflects the shape of the finished workpiece. For panel hollowing, the lower half of the guide will be concave, as shown here. For routing convex panels, the lower guide should be convex. Cut the guides out using the bandsaw arc-sawing jig shown on p. 118. Make the lower guides at least 6 in. longer than the width of the workpiece.

Attach the carrier frame and workpiece to a plywood baseplate. Screw a flat workpiece directly to the baseplate, locating the screws out of the router bit's path at the ends or edges. (Use brass screws, just in case.) For coopered or otherwise curved pieces, screw wood support strips to the baseplate (E), and then screw the panel down through countersink holes (F).

Place the router track onto the curved guides in the carrier frame. Outfit your router with a large core-box bit for a concave shape, or a big straight bit for convex shapes. Align the track with one edge of the work, and adjust the bit to cut no deeper than about $\frac{1}{16}$ in. Slide the router along the track to take a pass, and then move the track over a little and take another pass (G). (Waxing the router's sub-base will help it glide more smoothly.) Repeat the process across the entire workpiece, and then increase the cutting depth for another set of passes. Repeat as necessary. Once the shape is complete, clean up the remaining ridges (H) by hand or with a random orbit sander.

> ⚠ **WARNING** **When using a bit larger than the opening in the router base, set an upward travel stop to avoid raising the bit up into the base.**

E

F

G

H

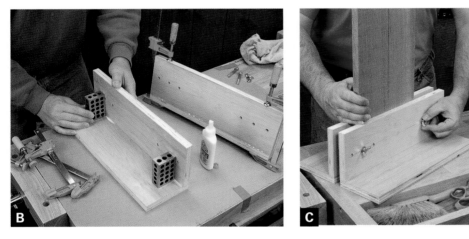

End-Routing Jig

Routing joinery, grooves, or decorative details into the ends of a board or frame member is practically impossible without an end-routing jig to support and align the router. To make a jig that'll handle stock up to 13 in. wide, cut four strips of ¾-in.-thick plywood to 5 in. wide and 16 in. long. Make sure that the long edges of these strips are perfectly square. Take two of the strips, align them perfectly on top of one another, and temporarily nail them together by driving small brads into the corners. Now take this block to the drill press and bore two sets of three ⁵⁄₁₆-in.-dia. holes centered along the length of the strip **(A)**. Locate the holes in each set 1 in., 2½ in., and 4 in. in from the end, and then pry the pieces apart and remove the brads. Next, glue the drilled strips to the others to form two L-shaped assemblies, using a pair of 1-2-3 blocks to assure that they're absolutely square **(B)**.

To use the jig, assemble the two halves with long bolts, washers, and wing nuts, choosing a pair of holes that allow bolting close to the edges of the workpiece. Place the jig upside down on a flat bench or machine table, and insert the work vertically, letting its end rest against the bench as you tighten the wing nuts **(C)**. Clamp the work into a bench vise and you're ready to rout, using an edge guide against one edge of the jig to guide the cut **(D)**.

Fluting Jig

Here's a jig for plunge-routing flutes or reeds along the length of a turned spindle, say for a table or chair leg. The jig is made up of two U-shaped assemblies that screw together, as shown in the drawing at right. The upper assembly consists of a router baseplate and two vertical support panels. This assembly carries the router over the work-piece and screws to the lower assembly, which slides along the lathe bed, guided by a runner that travels between the flat, twin rails found on most wood lathes.

Start by cutting out a square router plate from ¼-in.-thick tempered hardboard that's as long and wide as your router's sub-base. (Let's call this measurement X.) Mount this plate to your router after drilling the necessary mounting holes. Next, make the U-shaped lower assembly from three pieces of ¾-in. plywood. Cut its two side pieces 3 in. wide by X long, and its bottom piece X by X, minus 1½ in. on its width. Join the pieces with nails and glue so that the sides sit atop the bottom piece, aligned at opposite edges **(A)**.

Cut two ¾-in.-thick plywood sides for the upper assembly, making each one X wide and 1 in. to 2 in. longer than the distance from your lathe's bed to its turning axis. Saw out a "window" in each side for visibility during routing **(B)**. Clamp these sides to the sides of the lower assembly, and glue and screw the router plate on top **(C)**, adding four short glue

(Text continued on p. 158.)

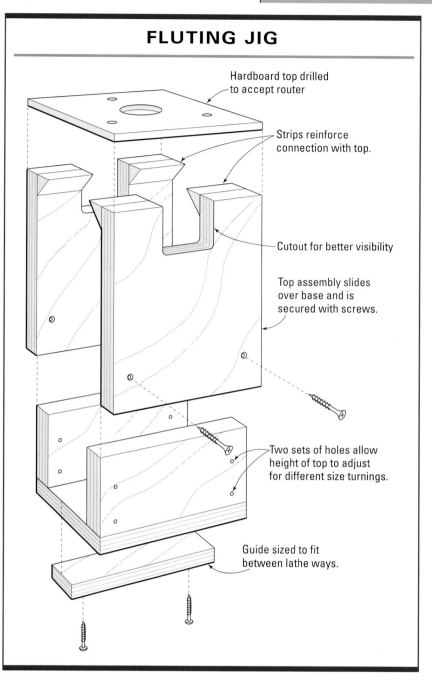

FLUTING JIG

Hardboard top drilled to accept router

Strips reinforce connection with top.

Cutout for better visibility

Top assembly slides over base and is secured with screws.

Two sets of holes allow height of top to adjust for different size turnings.

Guide sized to fit between lathe ways.

A

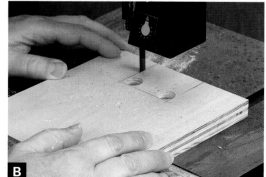

B

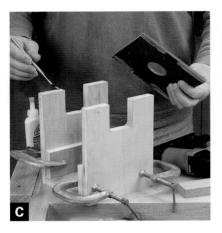

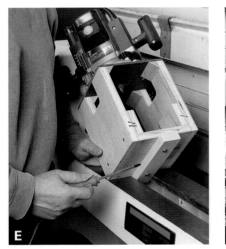

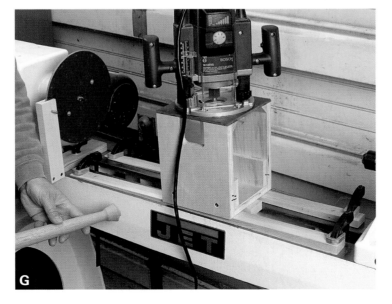

strips to reinforce the joint. With the units still clamped, drill a pair of holes near the bottom of each side into the base. Then remove the clamps, raise the sides up on a couple of 1½-in.-thick scraps, and drill another set of holes through the sides (D). The holes allow you to set the height of the router at a higher or lower position, to suit either small- or large-diameter turnings.

Make the runner that guides the carriage, ripping a 1-in.-thick hardwood strip wide enough to fit snugly between the lathe's rails without binding. Screw it to the underside of the base, making sure to align the router collet centerline along the lathe's spindle centerline (E). Before routing, prevent the workpiece from turning by securing it with the lathe's built-in spindle lock or a shopmade index plate.

With a bullnose or corebox bit chucked in the router, plunge the bit into the turning, and then slide the carriage along the lathe bed (F). For routing tapered spindles, clamp a pair of tapered wood shims (cut to match the taper angle of the turning) to the lathe bed (G).

[TIP] When routing multiple flutes or reeds, you'll get more uniform results by clamping stops to your lathe's bed to limit carriage travel.

Pattern Routing Jig

Imagine routing elaborate designs for signs or decorative cabinet parts using an illuminated pointer as a guide. With this jig, you rout the workpiece below a clear-plastic tabletop that holds your drawing while you're tracing along the drawing with point of light generated by a flashlight **(A)**.

Build the table apron and legs from 2½-in.-wide, ½-in.-thick plywood strips, making the table whatever size you like. Butt-join the apron pieces together with glue and nails, assembling them around a ⅛-in.-thick clear acrylic plastic top. Tack narrow ledger strips all the way around the inside of the apron to support the plastic **(B)**. Cut plywood strips for the table legs, making them about 2 in. longer than the height of your plunge router. Construct each leg from two strips joined to form an L shape **(C)**. Attach the legs to the apron with small clamps, so you can change table height to suit work thickness.

Cut a small MDF block to hold the pointer, and drill a hole halfway through it, sized to fit the head of a small LED flashlight (available at hardware and sporting goods stores). Although it may be tempting, DO NOT use a laser pointer, as there's a risk of eye injury. Without unclamping the block, finish drilling through it with a ⅟₁₆-in.-dia. bit to create a pinpoint beam **(D)**. Install the flashlight in the hole using a plastic tie or hot-melt glue. To support the

(Text continued on p. 160.)

E

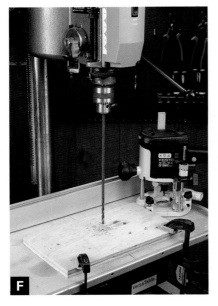

F

G

block, make a vertical wood rod that's slightly taller than your router and attach it to a non-plunging part of the base (E).

Next, the light pointer needs to be aligned with the axis of the router bit. To do this, clamp a plywood scrap to your drill-press table and lower the table enough for your router to clear the chuck by a few inches. Using a long, ¼-in.-dia. drill bit (a bellhanger's bit or a regular twist bit with an extension), bore a "guide" hole into the plywood (F). Install a ¹⁄₁₆-in. bit in the chuck, and mount a ¼-in.-dia. straight bit into your router. Plunge the router bit down slightly and insert it into the guide hole. Align the drill bit with the pinhole in the pointer block, lowering the bit into the pinhole. With the bit and pinhole aligned, glue the block to the end of the rod (G), using a few drops of thin cyano-acrylate glue. Clamp the jig and workpiece to a benchtop and adjust the table's height so the pointer block is ¹⁄₁₆ in. below the top. Tape your drawing to the top, switch on the flashlight, and start routing (H).

H

Four-Sided Base

Each edge of this offset replacement sub-base sits at a different distance from the bit. Therefore, when you're routing with a straightedge guide, the distance of the cut from the guide will depend on which of the four edges rides against it **(A)**. The sub-base is handy for quickly routing several grooves, dadoes, or slots at fixed intervals. Or, by making the dimensions of the four sides only slightly different, you can rout wide grooves by taking multiple passes with a single bit rather than having to reset the fence.

To make the sub-base, first calculate the desired distance of each side from the bit, taking the bit's radius into account. For example, let's say you want to rout a ½-in.-wide groove 6¾ in. from the edge of a board, using a ½-in.-dia. bit. The distance from the edge of the guide to the center will be 6¾ in. plus the ¼-in. radius of the bit, totaling 7 in.

Mark out all your distances on a piece of ¼-in.-thick hardboard **(B)**. Selecting the center-to-edge distance for the first two adjacent edges will determine the guide's center point. Marking out the distances for the two remaining edges will determine the trim lines for the base's final size. Cut the sub-base out on the tablesaw **(C)** and then check the dimensions with a rule and trim again if necessary. Now remove your router's stock sub-base and use it as a template for drilling mounting holes through the hardboard sub-base, taking great care to center the stock sub-base before drilling.

▶ See *"A Router Sub-Base Jig"* on p. 140.

By attaching a runner whose width matches the diameter of the bit, you can rout evenly spaced dadoes or grooves **(D)**.

Self-Centering Base

This router guide centers mortises and slots on the edges of frame members. Its twin fences allow routing right up to the end of the workpiece (A). The base is made from a piece of ¼-in.-thick hardboard cut to 6¾ in. x 9½ in. Mark the center point, and then draw a line through it across the width of the piece. Mark points on the line 1¼ in. on either side of center. With a compass leg on one of the points, draw a 4⅝-in.-radius edge-to-edge across the width of the piece on one end. Then repeat the procedure, swinging the compass from the opposite point. Bore two ¼-in.-dia. countersunk holes at those points for attaching the fences and then bore a centered hole for the router bit. Mount the base to your router, as shown on p. 140, and then bandsaw the ends of the base along the radius lines (B).

Make two ¾-in. x 1¼-in. x 12-in. solid-wood fences, and bore a ¼-in.-dia. bolt hole through the thicker dimension of each piece, centered along its length. Next, install a ¼-in. hanger bolt 1¾₆ in. from one end of each piece, adding fender washers and small hand knobs. Glue a small hardboard shim to the end of each fence next to the hanger bolt, and attach each fence to the base with a ¼-in. x 2-in.-long flat-head machine screw (C).

To set the guide, place it over the workpiece and pivot the fences until they're in full contact with the workpiece (D), and then lock the hand knobs.

[TIP] **You can rout off-center mortises by shimming the inner face of one of the fences.**

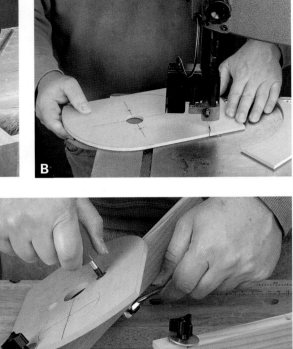

Angled Base

If you own a round-bodied router or laminate trimmer that removes from its base (e.g., Porter-Cable® 690 and 310 models), here's a neat carriage jig that tilts the bit at either 30 or 45 degrees relative to the work, for angled routing tasks.

The body of the jig is sawn from a square block made of laminated layers of ¾-in.-thick, high-quality plywood such as Baltic birch or maple die board. The cube should be 1 in. to 1¾ in. larger than the diameter of your router's motor housing. After making the block, draw a circle on one end of it, about ⅟₃₂ in. larger in diameter than the motor housing. Saw out a cylinder of waste on the bandsaw, starting the cut in the center of one side of the cube **(A)**.

Check the fit of the router motor and sand the contour if it binds. On the tablesaw, bevel one of the open ends of the cube at 45 degrees, and the opposite end at 30 degrees **(B)**. Also cut two fence strips from solid stock with beveled edges that match the cube cuts. Glue these strips to the sides of the block opposite the respective bevel cuts **(C)**. Use tape to hold the strips in place until the glue dries, and then drive a couple of brads to secure each one. Just above each fence strip, on the kerfed side of the cube, drill a hole two-thirds of the way through and install a screw to lock the router motor into the jig. Installing the router into one end of the jig produces a 45-degree tilt, while installing it on the other end tilts the bit for a 30-degree cut **(D)**.

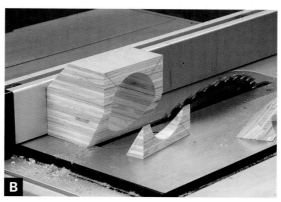

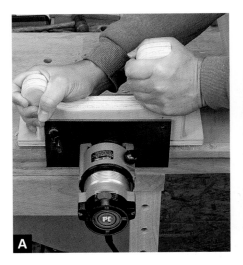

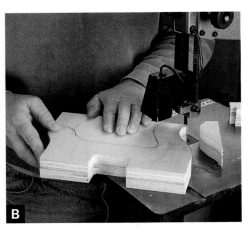

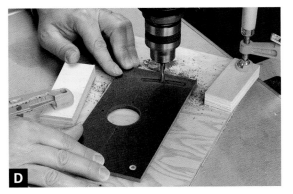

Edge-Routing Base

Mounting a laminate trimmer horizontally in an edge-routing jig allows cutting of slots and rabbets with straight bits. This jig's plane-like handle, which is glued up from two 7-in. x 10-in. layers of ¾-in.-thick plywood, makes the tool easy to control **(A)**.

Use a bandsaw or jigsaw to shape the handle, cutting out the bottom to allow chip clearance **(B)**. Glue and screw the handle to a 4-in. x 12-in. base made from ¾-in.-thick plywood. Notch the base for bit clearance, and rabbet its lower edge to allow edge-trimming **(C)**. Make the router mounting plate from a 4-in. x 8-in. piece of ¼-in.-thick tempered hardboard that you've drilled to attach to your laminate trimmer base. A slot and pivot hole in the plate are used to mount the plate to the jig and provide bit height adjustment. After drilling a hole for the pivot screw in the right upper corner of the plate, rout the curved slot on the drill press. To do this, screw the plate down through the pivot hole and use a ¼-in.-dia. spiral-fluted straight router bit to cut the slot, lowering the bit into the work and pivoting the plate **(D)**. Stops clamped in place limit the pivot motion to 2½ in.

On the side of the handle, drill a series of holes for the base pivot screw and install a ¼-in. threaded insert to accept the compass-locking handscrew **(E)**. After mounting the laminate trimmer to the plate, screw the plate to the handle, choosing a pivot hole that places the plate at the appropriate height for the job at hand. The handscrew locks the plate in the desired location.

Shooting Boards

A shooting board is used to guide a handplane when you're trimming the ends of mitered frame members and other parts. The board's sturdy twin fences support stock accurately at either a 45- or 90-degree angle as the handplane (lying on its side and sliding along a straight guide) is pushed past, trimming the part to a precise angle.

Start by cutting out a ¾-in. x 10-in. x 24-in. base and a ½-in. x 5-in. x 24-in. straight guide from MDF or plywood. Glue and nail the guide atop the base. Cut the 45-degree and 90-degree fences for the shooting board from kiln-dried solid stock at least 1½ in. thick. Avoid green construction lumber, which is likely to warp and crack as it dries. Cut a 5-in.-high, 45-45-90-degree triangle for the 45-degree fence **(A)**, and a 2-in. x 5-in. piece for the 90-degree fence. Make sure that the bearing edges of the fences are absolutely straight and square. Mount the fences to the straight guide with large wood screws set into countersunk holes **(B)**. It's best to screw the fences down and check their alignment with an accurate try square and miter square, and then glue and screw them back in place. Stick PSA-backed sandpaper to the face of each fence **(C)** to keep workpieces from slipping.

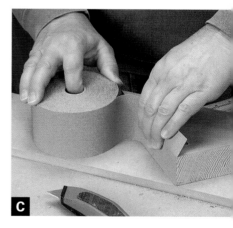

To use the shooting board, hold the workpiece firmly against the appropriate fence, with just a bit of the end overhanging the straight guide. Trim the end using a square-sided bench plane with a razor-sharp blade set for a very light cut **(D)**.

[TIP] For accurate shooting, make sure the sides of your bench plane are perfectly square relative to the sole.

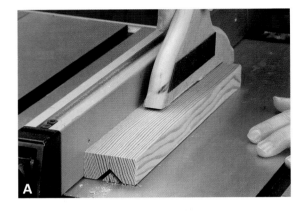

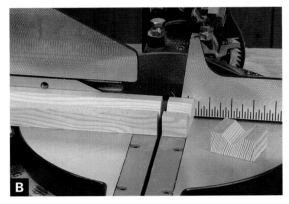

Small Crosscut Box

Sawing small moldings, strips, and dowels to accurate length is tricky, even when handcutting with a
regular miter box. This sawing jig lets you cut small
stock accurately and safely, leaving clean, square
ends. Start by cutting a V-groove in the face of a
1-in. x 2-in. piece of solid stock that's at least a foot
long **(A)**. Saw the 1⅛-in.-wide, ⁹⁄₁₆-in.-deep groove
using a tablesaw with its blade set at a
45-degree angle, making two passes to complete
the groove. Crosscut two short blocks off the end
of the strip: one 1 in. long, and one 1½ in. long **(B)**.
Use these blocks to create a sawing guide for a
fine-tooth, thin-kerf dovetail saw or Japanese
crosscut or flush-cut saw. Flip the blocks over and
glue them to the top of the strip, located about a
quarter of the way from one end of the V-strip **(C)**.
Clamp the blocks down with the sawblade between
them, as shown. After the glue dries, saw down
to the bottom of the V to complete the blade
guide slot.

Apply a length of self-stick measuring tape to one
face of the V-groove with its zero end perfectly
aligned with the sawblade slot. Finally, glue and
nail a square wood strip to the bottom of the jig to
allow for clamping the jig in a bench vise. To cut a
workpiece to precise length, align one end with the
desired measurement on the tape and hold it down
firmly as you saw through it **(D)**.

Templates

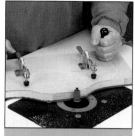

Shaping

➤ Pattern-Routing Templates (p. 170)
➤ Template Mounting Methods (p. 173)
➤ Baseplate Template Jigs (p. 174)
➤ Pin Routing (p. 175)

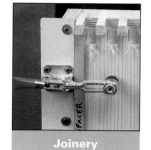

Joinery

➤ Mortising Templates (p. 176)
➤ Box Joints (p. 178)
➤ Complementary Routing (p. 180)

Drilling

➤ Drill Blocks (p. 181)
➤ Shelf-Pin Holes (p. 183)
➤ Pocket Holes (p. 184)

Sawing and Sanding

➤ Tablesaw Jig (p. 185)
➤ Bandsaw Curve-Cutting (p. 186)
➤ Drum Sanding (p. 187)

I F YOU'VE EVER made sheet cookies during the holidays, you've probably used cookie cutters to punch out tasty little Santas, trees, and snowmen. Templates are basically the cookie cutters of the woodshop. They can produce multiple parts that conform to an original pattern. Basic templates provide a guide for cutting, shaping, or trimming parts to final size, while more sophisticated templates create perfect inlays and sophisticated joinery.

The majority of template work involves edge-shaping operations performed with a router or shaper, using straight bits and cutters in conjunction with guide bearings, bushings, or pins in a process called pattern routing. But templates are also very useful for cutting tight-fitting joints such as mortise-and-tenons and box joints (see the photo on p. 168). Templates also excel when it comes to routing recesses and slots for mounting hardware, such as door hinges and locks. Apart from these shaping applications, there are template-routing methods that

With its built-in clamp and template base, this "clamshell-style" jig is useful for pattern routing dozens of identical parts on a shaper or router table.

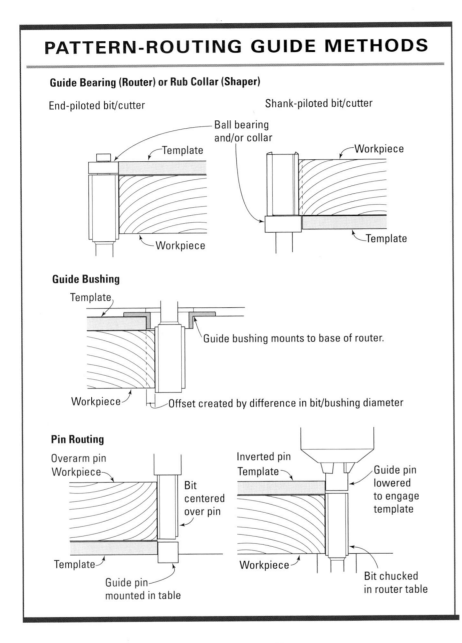

PATTERN-ROUTING GUIDE METHODS

Guide Bearing (Router) or Rub Collar (Shaper)

End-piloted bit/cutter

Shank-piloted bit/cutter

Ball bearing and/or collar

Template

Workpiece

Workpiece

Template

Guide Bushing

Template

Guide bushing mounts to base of router.

Workpiece

Offset created by difference in bit/bushing diameter

Pin Routing

Overarm pin
Workpiece

Bit centered over pin

Inverted pin
Template

Guide pin lowered to engage template

Template

Guide pin mounted in table

Workpiece

Bit chucked in router table

The fingers of this shopmade joinery jig guide a router bit that cuts a box joint—a strong way of joining the sides of boxes, chests, and drawers.

work with other machines, such as the tablesaw and drum sander, for sawing and sanding parts to shape.

Used with an electric drill or drill press, drilling templates bring speed and accuracy to common tasks such as boring angled holes or drilling rows of evenly spaced holes for adjustable shelf pins. Although templates are useful by themselves, this section will present several ways of incorporating them into various fixtures that serve to align and clamp parts in place for machining. For example, a pocket-hole jig incorporates a

► See "*Pocket Holes*" on p. 184.

drill guide into a fixture that holds the work at a tilt; screws driven into the steeply angled pocket holes create strong joints between frame members and cabinet members and cabinet parts.

Pilot Bearings and Guide Bushings

All pattern- and joinery-routing operations require that you guide a router bit (or shaper cutter) along the edge of a template. The two most popular methods use pilot bearings or guide bushings, as shown in the drawing at left. But choosing the one that's right for a particular application demands that you understand the pluses and minuses of each method.

A pilot bearing is a sealed ball bearing mounted directly onto a router bit (see the top photo on the facing page). The bearing guides the cut by rolling along the edge of a template or against the edge of the workpiece itself. The bearing on an *end-piloted* bit is screwed to a stud on the end of the bit, while the bearing on a *shank-piloted* bit is slipped over the shank and secured with a locking collar. (For shaper work, the bearing is called a "rub collar" and installs on the shaper's arbor, either above or below the cutter).

End-piloted router bits are terrific for flush-trimming and pattern-routing the outer contours of parts. Shank-piloted bits work for this application as well as for mortising and other interior routing operations. With a handheld router, the template is mounted above the workpiece for shank-piloted bits, or below the workpiece for end-piloted bits. (Template orientation is reversed when these same bits are used in a router table.)

Changing template orientation changes cutting direction and, depending on grain direction in the workpiece, can significantly affect the smoothness of the cut. (See photos I and J on p. 172.)

Commonly used "flush-trim"-style bits have bearings that match the diameter of the cutter, so they rout parts to exactly the same size as the template they're used with. But bits can also be fitted with bearings larger or smaller than the bit's cutting diameter, which can be useful when you wish to vary the size of the part relative to the template.

Unlike pilot bearings, guide bushings mount not to the bit but directly to the router's sub-base, (see the bottom photo at right) providing a solid guide that doesn't rotate. Guide bushings work in many of the same situations as shank-piloted bits. Bushings are available in a variety of different outside diameters, typically ranging from ¼ in. to 2 in. They're particularly helpful when used with plunge routers and templates for cutting mortises or box joints. The bushing stays in contact with the template as

Bearing-guided bits include flush-trim bits, end-piloted bits, and shank-piloted bits. Some have flush bearings, while others are larger or smaller than the bit's cutting diameter.

A metal guide bushing mounted in the base of a router provides an effective way of guiding a bit along a template to shape the edge of a part, to cut joinery, or to rout recesses for hardware or inlays.

the bit is lowered into the work. (You could do this with some shank-piloted bits, but you might slip and cut into the template while plunging.)

Because they must be used with the template between the workpiece and the router, guide bushings are less versatile than piloted bits. Furthermore, the bit diameter must be smaller than the inside diameter of the bushing, requiring compensation in the template for the bit/bushing offset.

CALCULATING TEMPLATE OFFSET

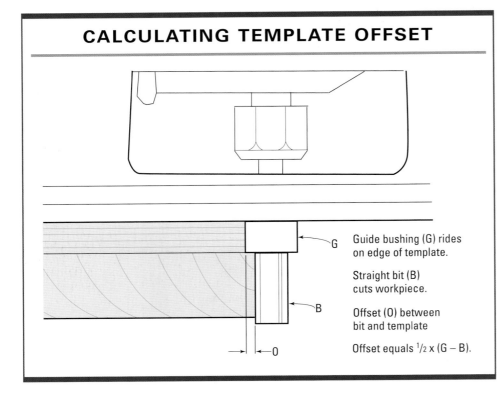

Guide bushing (G) rides on edge of template.

Straight bit (B) cuts workpiece.

Offset (O) between bit and template

Offset equals $\frac{1}{2} \times (G - B)$.

Pattern-Routing Templates

Pattern routing is an easy and powerful way to create any number of shapely parts out of wood, from furniture legs to toy ducks or cheese boards. The method works much the same way as any routing or shaping operation in which a bearing-guided bit or rub collar follows the edge of the workpiece **(A)**. In pattern routing, the bit or collar follows the edge of a shaped template attached to the workpiece, instead of the edge of the workpiece itself.

Templates can be cut from practically any rigid material that has a hard edge, including MDF, tempered hardboard (Masonite), Baltic birch plywood, and many plastics. The thickness of the templates can range from ¼ in. to ¾ in. Generally, templates used with end-piloted bits should be thicker, while bushing and shank-piloted bits call for the use of thinner templates in order to get the maximum depth of cut.

The first step in template making is to create a full-scale pattern of the desired part on a piece of paper or thin cardboard. The edges of cardboard template patterns cut out with scissors are easy to smooth and even up using a drum or oscillating-spindle sander. Remember that the radius of the pilot bearing, guide bushing, or guide pin that you plan to use should be smaller than the smallest inside radius of any curve on the template. Otherwise the guide won't accurately follow the template.

Templates used with a flush-trim bit are easy to make because the template is exactly the same size and shape as the part. But templates made to use with a bearing or bushing that differs in diameter from the bit **(B)** require more planning, because you must compensate for the offset of the bit/bearing combination if the resulting parts are to come out as desired. As shown in the drawing above, the general formula for calculat-

ing offset is to take the outside diameter of the pilot bearing (or guide bushing, rub collar, or guide pin), subtract the cutting diameter of the bit, and divide by two.

The formula holds true whether a template is for shaping parts or for cutting inlays or joinery. Hence, if you were routing one edge of the work using a ½-in.-dia. straight bit and a ¾-in. O.D. guide bushing **(C)**, you'd end up a part that's ⅛ in. wider than the template due to the ⅛ in. offset between the bit and the bushing. Also, remember that when you rout a circular recess or rout all the way around a part, the offset is doubled.

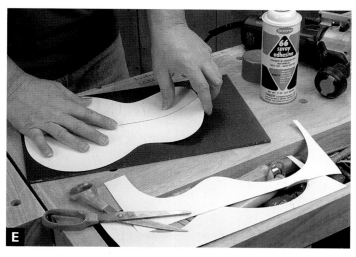

Making an offset template for a simple shape like a rectangular mortise is very straightforward: Just shift the template's edge over from the desired pattern line by the amount of the offset. But to create an irregular template with an offset, follow the edge of the pattern with a draftsman's compass set to the desired amount of offset **(D)**. If the pattern is curved, take care to keep the compass/dividers perpendicular to the line, or you'll distort the design. Complicated patterns are much easier to redraw for correct offset, using a photocopy machine with variable reduction/enlargement. Check the photocopy by overlaying the original pattern and measuring the amount of offset.

Once you're ready to cut the template out, attach the pattern directly to the template stock using spray adhesive **(E)** (available from an art supply store). Clean up the edges of the sawn template by hand or with a power sander, making sure that all straight edges are straight and curved edges flow smoothly **(F)**. Any irregularities will be re-created in shaped parts, making for more cleanup work later on. When shaping the single edge of a part such as a chair leg, extend the edge of the template past where the cut begins and ends, to give the guide bearing/bushing a smooth start and finish.

(Text continued on p. 172.)

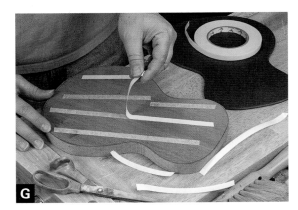

Rough-cut the workpiece to approximate shape so the bit/cutter doesn't need to hog off tons of material. Then attach the template using double-stick tape, small nails, or other methods (see "Template Mounting Method" on the facing page) **(G)**. For thick or heavy parts, screws hold better.

Parts large enough to clamp flat atop a workbench are most easily routed freehand **(H)**. For smaller parts, it's best to use a router table or shaper and, if possible, to incorporate the template into a baseplate jig.

▶ See *"Baseplate Template Jigs"* on p. 174.

The template/workpiece is oriented either facing up or down, depending on whether the guide bearing is above or below the cutter. When template-routing curvy parts and splintery woods, you'll get cleaner edges with less tearout if you use a flush-trim bit fitted with two pilot bearings, one on the end, and one on the shank **(I)**. This allows you to flip the template/workpiece over, and rout favoring the grain direction as it changes in different sections **(J)**.

[TIP] To template-rout two parts at once, use thick stock and then resaw it in half after routing.

Template Mounting Methods

One of the quickest ways to secure a template to the work is to use an adhesive transfer tape gun **(A)**. This clever device uses a trigger-controlled rolling motion to put down a coating of the same kind of sticky stuff found on double-stick tape. After the template and workpiece are pounded together with a mallet, the bond is strong enough for all but the heaviest shaping work. The adhesive rubs off easily after shaping.

An even more elegant method to hold the template to the work is to use the force of a vacuum **(B)**. A vacuum template, made from plastic or another nonporous material, connects via a small plastic hose to a vacuum pump or special vacuum valve that creates enough suction to keep it firmly planted during shaping.

➤ See *"Vacuum–Clamp Systems"* on p. 218.

A

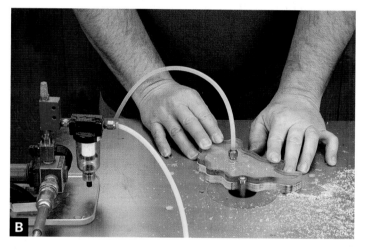

B

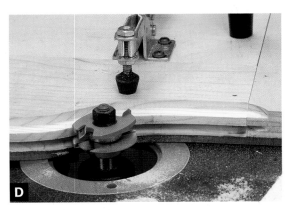

Baseplate Template Jigs

What's easier than shaping a part with a router and template? Using a jig that incorporates the template as part of a jig that positions and clamps the workpiece atop the template with no other fastening required. Baseplate template jigs used with a router table **(A)** or shaper are so efficient that they're a mainstay for production woodworkers. The only limitation is that you can't use a baseplate template for shaping parts all the way around.

Start by cutting out a template for the desired part shape, as described previously. Make the template at least a couple of inches wider than the part to leave room for the stops and clamps. Extending the edges of the template a little at both ends gives the bearing or collar support at the start and end of cutting, resulting in a cleaner-cut part. Attach stop blocks that were cut from stock the same thickness as the workpiece to the top of the template, driving nails or screws in from underneath **(B)**. Locate the stops so they'll properly position the work blank relative to the shaped template edge.

To hold the part firmly, you can use eccentric clamps, cam clamps, or quick-action clamps **(C)**. For smaller jigs, adding a holding plate to the clamp head and covering the surfaces with PSA sandpaper helps the part stay put during shaping (see the bottom photo on p. 167). On larger jigs, add a pair of handles for better safety and control. With the rough-trimmed part clamped in place, flush-trim the edge of the part using a bottom-piloted bit or cutter. By changing bits/cutters, you can also use the baseplate template jig to further shape the part's edge, perhaps to rout a curved cabinet rail for cope-and-stick joinery, for example **(D)**.

[TIP] **When template routing, save time by stacking blanks and sawing out several rough-cut parts at the same time on the bandsaw.**

Pin Routing

If you don't have the luxury of owning an over-arm pin router, inverted pin routing can provide a practical alternative. In lieu of buying a special overarm attachment for the router table **(A)**, you can do inverted pin routing with a regular router and a drill press, mounting the guide pin above the bit in the drill press's chuck **(B)**. The router is housed in a small router table **(C)**, which is basically a plywood box with an open front and back for access to the router. The screwed-on tempered hardboard top is drilled for hanging the router underneath.

To use the setup, position the router table atop your drill-press table and carefully align a straight bit mounted in the router with a guide pin (a metal rod the same diameter as the bit) mounted into the drill-press chuck **(D)**. In preparation for routing, a template of the desired size and shape is fastened or clamped to the top of the workpiece **(E)**.

In this example, a rectangular template is being used for routing out the insides of a small wood box. To make routing easier, most of the inside waste is first bored out with a large drill bit **(F)**. After setting the router's bit height, place the workpiece over the bit (making sure it can rotate freely) and lower the drill chuck until the pin engages the template. After switching on the router, move the part carefully over the bit, allowing the guide pin to gently contact the edges of the template at the perimeter of the cut. For inside cuts as shown here, you must occasionally stop the cut and clean out the waste trapped inside the workpiece.

> ⚠️ **WARNING** When using a pin-routing setup with a router table, always make sure the bit is clear of the stock when you're switching on the router.

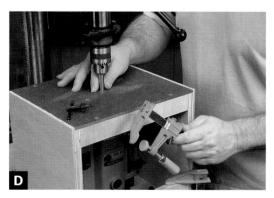

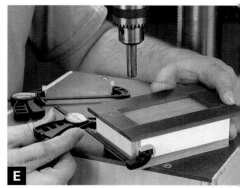

Mortising Templates

Rectangular templates are easy to make and are great for cutting mortises for traditional mortise-and-tenon joints or for loose tenon joints. The template is mounted atop a fixture that positions and holds the workpiece firmly while it's routed **(A)**. Instead of trying to cut out a perfectly rectangular hole, it's simpler to make a mortise template by assembling it from several separate strips of MDF or tempered hardboard.

Start by calculating the exact size of the template opening. If you're using a shank-piloted bit with a bearing the same diameter as the cutter, make the opening the same size as the desired mortise. If a straight bit and guide bushing will be used instead, compensate for the offset between the bit and bushing.

► See *"Pilot Bearings and Guide Bushings"* on p. 168.

For example: with an ⅛-in. bit/bushing offset, the template opening for a 1-in. x 4-in. mortise would be 1¼ in. x 4¼ in.

Cut two strips for the ends of the template, making them as wide as the desired opening. Then cut two side pieces, making them wide enough to provide a stable platform for your router. You can get away with making small mortise templates from ¼-in.-thick stock, using the gluing method described earlier.

► See *"Mortise Template"* on p. 54.

Large mortise templates (or ones that will get lots of use) are best made from stout ½-in.-thick MDF, using plate joinery biscuits to reinforce the connections between parts **(B)**.

After applying glue and clamping up the parts, measure the exact template opening and slide parts in or out as necessary **(C)**. After the glue

dries, trim the template ends square and flush. You can clamp the template directly atop the work, mount it to a simple, box-like fixture that clamps to the workpiece **(D)**, or mount it to a more elaborate clamping fixture, like the one shown in the top photo on the facing page. You can either square up the rounded corners of routed mortises with a chisel, round over the tenon edges, or use loose splines.

➤ See *"Routing Fixtures"* on p. 222.

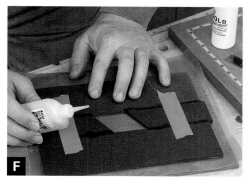

A router and template are also great for creating recesses or sockets for inlays and hardware. A small template with a rectangular opening is very handy for routing shallow mortises for butt hinges **(E)**. Gluing a pair of stop strips to the bottom of a hinge template will align and space the template so the hinge recess is routed in the right place on the workpiece. The strips also provide a handy way of clamping the template to the work.

Inlay templates may be practically any shape: polygonal, curved, or irregular. For symmetrical forms, you may employ the same method used for making rectangular templates. The diamond-shaped inlay as shown here **(F)** has 60-degree mitered angles cut on the two pieces that make up the ends of the template. Clamp the template atop the work surface and rout, using a shank-piloted or guide-bushing-guided straight bit **(G)**.

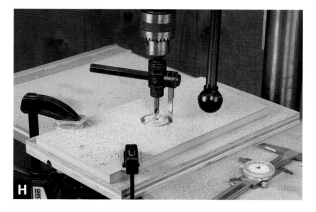

You can also use template routing to cut the 35mm sockets required for mounting European-style cabinet hinges. After calculating bit offset, use an adjustable fly cutter to bore a hole in the template stock **(H)**. Add a stop strip to space the hole relative to the edge of the work, and rout as you would for any other mortise **(I)**.

[TIP] If necessary, shim the edges of a template with masking tape to reduce the dimensions of a routed mortise or inlay.

BOX JOINT JIG

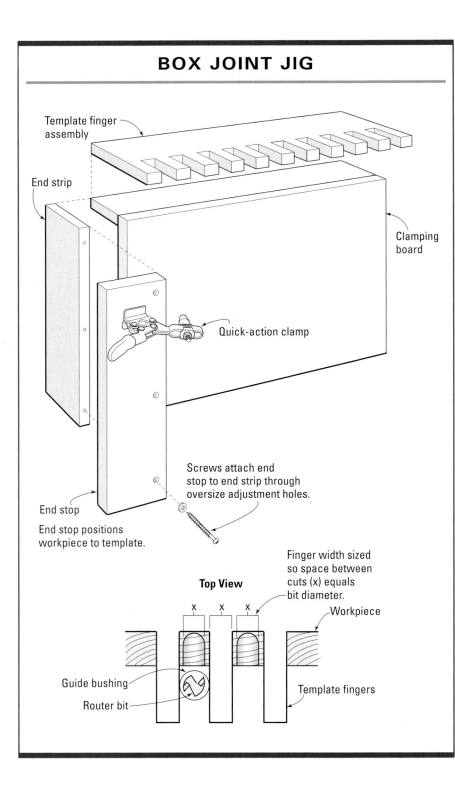

Template finger assembly

End strip

Clamping board

Quick-action clamp

Screws attach end stop to end strip through oversize adjustment holes.

End stop

End stop positions workpiece to template.

Top View

Finger width sized so space between cuts (x) equals bit diameter.

Workpiece

x   x   x

Guide bushing

Template fingers

Router bit

Box Joints

As the name implies, finger templates have many identical, evenly spaced fingers that are used to guide a router for cutting a box joint in stock that is clamped to the fixture beneath the template. While a tablesaw box joint jig is great for cutting shorter parts, a finger template setup can rout workpieces of practically any length.

▶ See *"Box Joints"* on p. 101.

The template jig shown in the drawing at left is designed to cut ½-in.-wide box joint fingers in stock up to 12½ in. wide and 1 in. thick. It is assembled from 14 fingers and 13 spacers cut from ½-in.-thick Baltic birch plywood and glued together edge-to-edge. These box joints are cut using a ½-in.-dia. straight bit guided by a ⅝-in. O.D. guide bushing on the router. Therefore, the spacers are cut exactly ⅝ in. wide, and the fingers are ⅜ in. wide.

Start by ripping long strips of plywood to a precise width for the spacers and fingers (each strip should measure exactly 1 in. wide) **(A)**. Then crosscut the strips into 4¼-in.-long fingers and 2¼-in.-long spacers. Make an alignment form for glue-up by tacking a couple of strips to a piece of scrap plywood covered with wax paper. Brush glue on the parts and then load them into the form **(B)**. Clamp a straight strip of wood over the fingers to keep them flat as you clamp them together **(C)**.

Make the jig's clamping fixture from two butt-joined pieces of ¾-in.-thick plywood: a 1¾-in. x 13⅜-in. top strip, and a 10-in. x 13⅜-in. vertical clamping board. Scrape off the excess dried glue on the template and then fasten it atop the fixture **(D)**, aligning it so that the fingers protrude 1½ in. past the face of the clamping board. Nail and glue a 2½-in.-wide end strip to the left side of the fixture, as shown in the drawing on the facing page. Screw the end stop—a 1¾-in. x 10½-in. strip of ½-in.-thick plywood—to the front of the end strip using washer-head screws driven into over-sized holes that allow a little side-to-side adjust-ment. Position the stop so that its edge is 7⁄16 in. from the inside edge of the first finger and per-fectly square to the template. To clamp parts more quickly and easily, mount a quick-action clamp to the end stop, positioning it far enough below the template to clear the router bit.

Secure the template fixture to your workbench by clamping the end strip into a bench vise. Clamp the first workpiece into the jig, making sure it's set flush to the end stop and against the underside of the template. Set the router bit to cut very slightly deeper than the thickness of the workpiece, and slip the guide bushing in between the first set of fingers to take a cut. Work your way along the template to complete cutting the first half of the joint **(E)**. To make the other half of the joint, offset the mating work-piece relative to the template by positioning a ½-in.-thick spacer against the end stop **(F)**. Rout as before, and your box joint is ready to assem-ble. You can quickly cut box joints in stock up to ½ in. thick by routing both pieces at the same time. Simply clamp both workpieces into the jig, one on top of the other, offset by the spacer strip as shown **(G)**.

[TIP] **Check the fit of joints produced by a finger template jig before gluing it up: Dry-clamp the template and carefully rout a test part.**

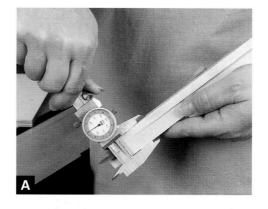

A

B

C

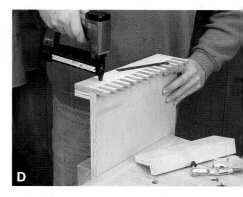

D

E

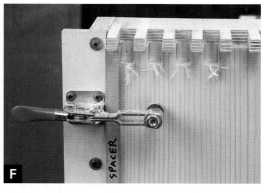

F

G

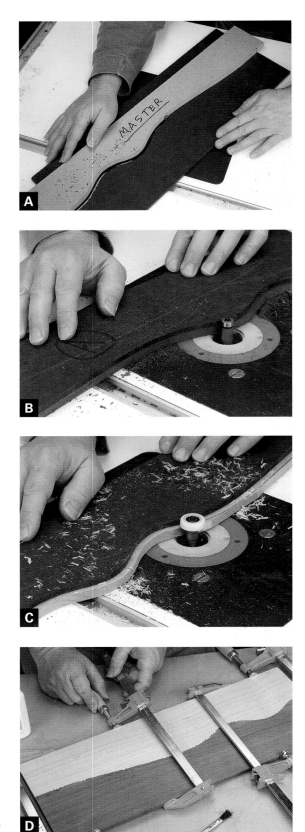

Complementary Routing

Want to join two parts together along a curva-ceous or irregular line for a decorative joint? The complementary template method uses a pair of shopmade templates, a router, and three piloted bits to produce parts that fit together perfectly. Create a "master template" first by drawing the desired joint line on a piece of MDF or hardboard (to work with the bits used here, none of the curves should have a radius of less than ¾ in.). Cut out and sand the template edges smooth and regular.

Use double-stick or adhesive transfer tape to temporarily affix the master template atop a piece of ¼-in.-thick hardboard that's twice as wide as the master, centering the joint line down the middle. Use a ¼-in.-dia. flush-trim bit in the router table to cut the hardboard in half by carefully run-ning the bit's pilot bearing against the master template **(A)**. This creates a pair of complementary templates. Label the one under the master tem-plate A and the other one B, and always keep these sides facing up during the following steps.

Affix template A to the top of a rough-cut work-piece and rout its edge to shape using a flush-trim bit **(B)**. Mount template B to the top of a second rough-cut workpiece, and rout its edge using a ½-in.-dia. straight bit fitted with an end-mounted 1-in.-dia. pilot bearing **(C)**. The ¼-in. offset between this bit and its bearing creates a contour on the second workpiece that exactly matches the flush-trimmed edge of the first! The joint is ready to be glued together without any further sanding of the routed edges **(D)**.

[TIP] You can use the complementary template method to rout both a recess and an inlay that fits snugly into the recess.

Drill Blocks

Drill blocks help you drill holes square to a surface or at an accurate angle without having to resort to using a drill press, which may not work with large workpieces anyway.

The simplest way to align a bit used freehand with an electric drill is with a V-block. You can make this kind of drill block by either sawing a V-shaped slice out of a hunk of 2x4 lumber, or by gluing together two thinner blocks, each with a 45-degree bevel on their ends. (The non-pointy side of each bevel faces inward, creating a V-groove where they meet.) To keep bits square, hold the block flat on the work surface and press it lightly against the bit so the bit is cradled by the V-groove **(A)**. The beauty of this simple jig is that it works with cylindrical bits of practically any diameter.

Another handy drilling jig is made by boring a hole of the desired diameter through a scrap block of wood. The hole then acts as a guide for bits of the same diameter when the block is held against the work **(B)**. Such drill blocks are great either for keeping bits square or for aligning them at a particular angle to the work. In fact, you can make a block serve double duty by cutting its opposite faces at different angles. In the block shown **(C)**, the drill is aligned at 90 or 22.5 degrees, depending on which face of the block is held against the workpiece. The limitation is that the jig is only good for the particular diameter of bit you make it for.

A hardwood block is durable enough to maintain drilling accuracy for dozens or even hundreds of holes. However, for greater accuracy and durability, you can fit the block with a bronze bushing **(D)**. Available from well-stocked hardware stores and woodworking supply catalogs, bronze bushings come with inside diameters that fit a wide assortment of standard bit diameters. Simply

(Text continued on p. 182.)

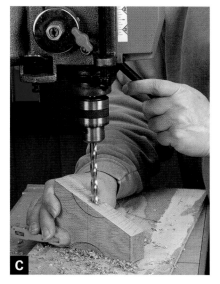

bore the hole in the drill block to fit the bushing's outside diameter, and drive it into the hole with a mallet.

You can also create drilling jigs to help position and drill multiple holes for joining carcase components, or for mounting cabinet hardware such as shelf guides, handles, and hinges. For example, you can make a jig for accurately orienting and drilling the two holes for a pull handle on a cabinet door. Start by drilling holes in a piece of plywood, spacing them the proper distance from each other and the desired distance from the door edges **(E)**. To use with two different-sized pulls, or for positioning pulls at different distances from the edge of the door, drill two or more sets of holes in the template. Glue and nail a couple of thin-wood stop strips to the edges of the plywood to position the jig on either the left-hand or right-hand corner of the door (or drawer front) **(F)**.

Drill templates with multiple holes are also great to use for boring pilot holes for the screws that mount door hinges. Bore the holes to match both the pattern and spacing of screws on the particular hinge you're using. Since hinges are usually fitted in pairs, make two drill templates and attach them to a single stop strip that will space them properly relative to each other and to the edge of the work. For a more flexible arrangement, mount two or more hinge drilling templates to a T-track **(G)**. This allows you to change the spacing of the hinges to suit doors of different heights **(H)**.

Shelf-Pin Holes

Drilling holes for adjustable shelf pins can be a hassle. If all the holes aren't spaced and located precisely, you can end up with crooked or wobbly shelves. Used with a special ¼-in.-dia. self-centering bit with a spring-loaded sleeve, a shelf-pin hole template **(A)** helps you get perfect holes every time.

Make the template from a 4-in.-wide piece of ½-in.-thick MDF (or two pieces of ¼-in.-thick Masonite laminated together) that's at least 4 ft. long—longer if you're working with really tall cabinets. Drill a row of ⅜-in.-dia. holes in the template (for the self-centering bit's sleeve) that are all spaced exactly the same distance apart—anywhere from 1 in. to 2 in. You can do this by precisely marking out the holes and drilling them on a drill press, but it's much easier to use a drill indexing jig **(B)**.

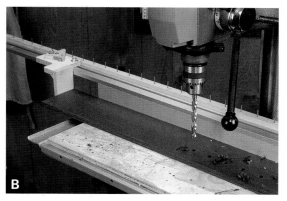

➤ See "Hole Spacing Jigs" on p. 202.

Space the row of holes 1½ in. from the template's edge to leave room for stop strips cut from ¾-in.-thick stock. Screwed to the underside of the template **(C)**, the stop strips butt against the workpiece, thus positioning the holes a fixed distance from the edge (typically 1½ in.–2 in.). Finally, draw a line at 90 degrees across the template, through both the first and last holes, to help align the template at the bottom (or top) of the workpiece.

If your template isn't long enough to drill all the shelf-pin holes you need in one go, then make a small positioning pin by turning the end of a ⅜-in.-dia. dowel down to ¼ in **(D)**. To reposition the template further along the length of the workpiece, slide the pin through the template and into the last hole you bored.

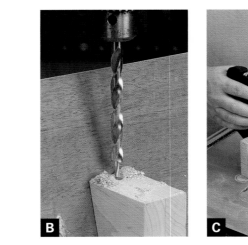

Pocket Holes

Pocket holes provide a handy way of screwing wood parts together. Steep-angled pocket holes are drilled in one of the parts (a cabinet side, face frame member, etc.), and then adjacent parts are butted and screwed together, making a strong, yet reversible connection. A pocket-hole jig **(A)** makes the whole process quick and easy. This jig mounts to the drill-press table and supports the stock at a 15-degree angle while a drill block guides a special stepped drill bit **(B)** so you get a clean pocket hole every time.

Make the drill block either from 8/4 stock or by gluing up two 2½-in. squares of 4/4 rock maple or other dense hardwood. Bore a ⅜-in.-dia. hole 90 degrees through the block, centering it edge-wise and positioning it about ¼ in. in from the corner. Set your tablesaw's miter gauge to 15 degrees, and cut a wedge off the edge of the drill block, slicing at an angle through the hole, as shown **(C)**.

Make the base of the jig from a 4½-in. x 20-in. piece of ¾-in.-thick wood or plywood. With a ¾-in.-wide dado blade tilted to a 15-degree angle, cut an angled groove into the base, so shallow that one side of the V barely penetrates the sur-face. Cut a 3½-in.-wide fence for the jig from ¾-in.-thick stock, beveling its lower edge at 15 degrees. Glue and nail the fence to the base so it aligns with the V-groove, and add wedge-shaped glue blocks to reinforce the back of the fence **(D)**. Center the drill block on the base, spacing it far enough from the fence to accom-modate the thickness of the workpiece **(E)**, and screw it on from underneath.

Tablesaw Jig

This template jig jig can help you saw out multiple straight-edged parts quickly and safely on the tablesaw, including triangles and polygonal shapes **(A)**. The setup uses a special overhanging fence that's clamped to the rip fence directly above the sawblade. A template atop the workpiece contacts the fence and guides the cut.

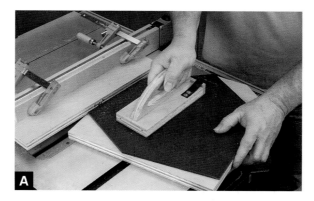

Make the fence from a couple of strips of scrap plywood glued and nailed into an L-shape. The vertical strip (the part that gets clamped to the rip fence) only needs to be about 2 in. wide and a foot long. The horizontal strip that contacts the template should be 16 in. to 20 in. long and wide enough to provide clearance for the portion of the workpiece that's cut off.

After unplugging the saw and raising the sawblade slightly higher than the thickness of the workpiece, center and clamp the overhanging fence to the rip fence so that the horizontal strip is parallel to the table and nearly touching the blade; use a pair of shim blocks to hold the fence as you clamp it **(B)**. Set the rip fence so the edge of the overhanging fence is flush with the left-hand blade teeth.

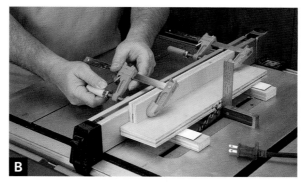

Cut a template from MDF or hardboard that's the exact size and shape as the desired part, attach it atop the workpiece with double-stick tape or adhesive-transfer tape, and hammer it down so it sticks firmly in place **(C)**. Switch on the saw and carefully guide the piece through the cut, keeping the template firmly in contact with the overhanging fence After each cut, shut off the saw and use a stick to push the scrap out from beneath the overhanging fence **(D)**, to keep it from accidentally binding against the blade on subsequent cuts.

WARNING Never template-saw parts with edges shorter than 4 in. long.

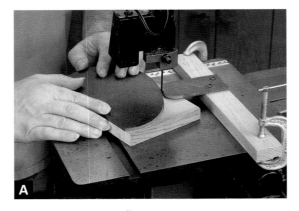

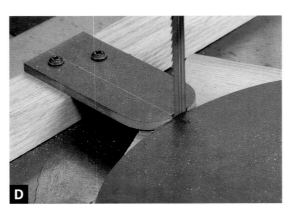

Bandsaw Curve-Cutting

Another type of template-sawing guide is useful for sawing out multiples of curved parts on a bandsaw. The jig employs a round-ended template guide clamped to the bandsaw **(A)**. A template temporarily mounted on top of the workpiece contacts the guide to regulate the cut. Make the guide from a 1½-in. by 4-in. piece of ¼-in.-thick tempered hardboard. Cut one end of the guide half-round, and then use a file or small rasp to make a notch in the center of the guide **(B)**. The notch should be slightly wider and deeper than the bandsaw blade (a ¼-in.-wide blade will work best). Screw the square end of the guide to a wood strip that's slightly thicker than the workpiece and long enough to span the length of your bandsaw's table **(C)**. Position the guide on the length of the strip so that the notch is even with the bandsaw blade. Clamp the guide to the table with the bandsaw blade seated in the notch, rubbing lightly against it.

Cut a template from thin hardboard that's the same shape as the desired part. The jig does allow you to cut concave as well as convex curves, but no concave curve should have a radius less than the radius of the guide's rounded end. Attach the template atop the workpiece and start cutting, keeping the template in firm contact with the guide as you proceed **(D)**. (If the waste portion of the work is too wide to fit under the guide, you may have to trim it off before template sawing.) With a little care, template-sawn curved parts can come out clean enough to require just a little sanding, either by hand or using the drum-sanding template setup described next.

Drum Sanding

When you want to sand multiple parts and keep them identical, template drum-sanding is the way to go. If you already have a drum-sanding dust table for your drill press, a special tool called a Robosander features a rotating guide bearing that allows accurate flush sanding to a template attached to the workpiece **(A)**.

> See *"Drum-Sanding Dust Table"* on p. 255.

Another way to template sand is to mount a disc, directly below the sanding drum, that contacts the workpiece template to guide the sanding process. Use a sanding drum that's slightly smaller in diameter than the smallest-radius concave curve on the workpiece. Cut the disc from MDF or hardboard that's slightly thinner than the template you're using.

For pattern sanding rough-cut parts, make the disc exactly the same diameter as the sanding drum. However, if you want to use this jig to finish-sand parts that have been template routed or sawn, make the guide disc slightly smaller than the drum—about $\frac{1}{16}$ in. or so less in diameter. Use the circle jig described on p. 116 to make a correctly sized disc that's perfectly round **(B)**. Screw the disc to the top of a baseplate that's large enough to fit to your drill press table. Use a pair of blocks to help center the disc under the sanding drum and clamp the baseplate down **(C)**.

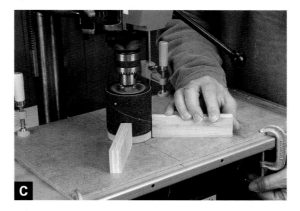

Lower the sanding drum until it nearly touches the disc, attach the template to the work, and you're ready to sand. Press the wood against the drum lightly at first, working your way around in several passes **(D)**, until the template is in full contact with the disc on the final pass.

[TIP] **Template sanding is a great way to clean up the edges of template-sawn or template-routed parts.**

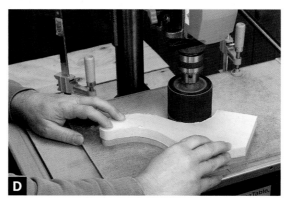

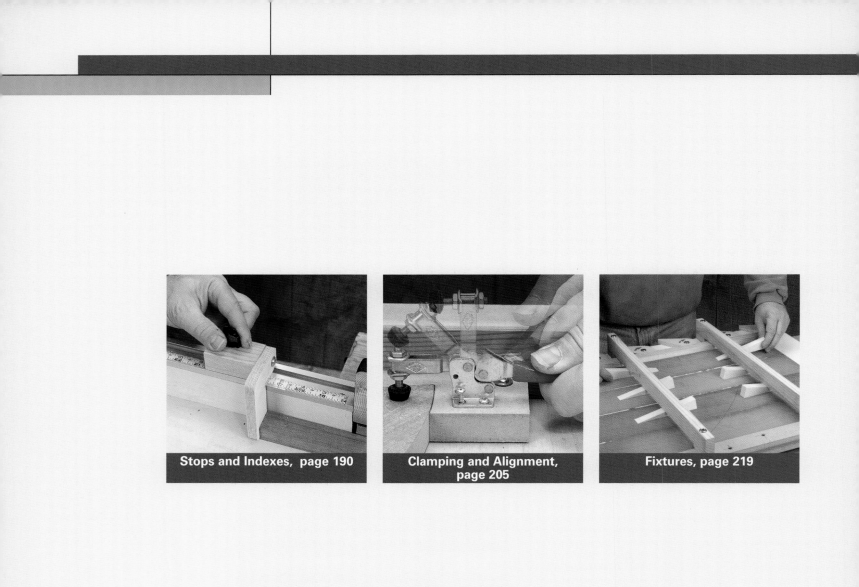

PART THREE

Woodworking Fixtures

WHAT'S THE DIFFERENCE between a jig and a fixture? Strictly speaking, a jig regulates the motion of a tool or the part that's being cut, while a fixture simply clamps a part or parts in place for machining or assembly. The simplest fixtures support workpieces atop machines, while elaborate fixtures may incorporate templates or forms that dictate the shape of parts to be cut or laminated.

Of course, you can't cut, plane, drill, rout, or sand wood safely without holding it firmly, so clamping and positioning devices are essential jig-building components. Stops and indexes orient stock for cutting or drilling at precise locations or regular intervals, while positioning blocks align parts for clamping. Shop-made wedges, cams, and other clamps secure work on a benchtop or in a jig to prevent a part from moving as it's cut.

Stops and Indexes

End Stops

➤ Flip Stops (p. 192)
➤ Extension Stops (p. 194)
➤ Eccentric Stops (p. 195)
➤ Fence Angle Stops (p. 196)

Travel Stops

➤ Depth Stops (p. 197)
➤ Travel Stops (p. 198)
➤ Routing Stops (p. 200)

Index Devices

➤ Hole-Spacing Jigs (p. 202)
➤ Kerf Spacing (p. 203)
➤ Lathe-Indexing Plate (p. 204)

WOODWORKERS STRIVE for accurate measurements when making project parts. Some measurements are regulated by tool settings, such as the depth of cut on a tablesaw or handplane. Other measurements are determined by our eyes and hands, such as when we're chopping a mortise to layout lines. Jigs and fixtures can also help us gauge measurements. When you're using jigs and fixtures, it's extremely useful to incorporate stops and indexes that can be set to determine the length or width of parts, the depth of routed slots and grooves, and the spacing of holes or flutes, among other things. Stops and indexes also serve to make accurate operations easily repeatable, which is very important when you're making

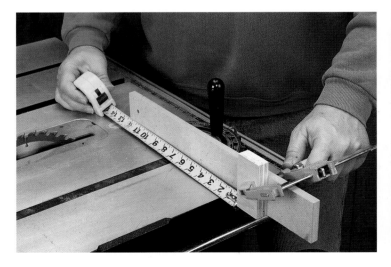

A block of wood clamped to a miter gauge fence determines the length of a part to be cut on the tablesaw.

These adjustable router stops clamp on the edge of the stock, limiting the travel of the router and hence, the shape routed on its edge.

a number of identical parts. As with other jig components, there are many kinds of stops and index devices to choose from—each appropriate for a particular range of tools and applications. A stop can be as simple as a block of wood clamped to a fence or table to set the length of cut, as shown in the bottom left photo on the facing page. More ingenious stop devices include flip-up end stops used for crosscutting, extension stops that expand the capacity of many different sliding jigs, travel stops that limit the length of dadoes and rabbets or edge-routed shapes, as shown in the bottom right photo on the facing page, and depth stops that determine how

deeply holes are bored. Still other stop devices are extremely handy for accurately and quickly setting the position of jig parts. For example, a micro-adjuster can be used to fine-tune the exact angle or position of a fence on a mitering jig or router table.

Index devices are specialized stops useful for setting the spacing of repeated machining operations. For example, you can use a hole index to consistently space shelf-pin holes in a bookcase side. You'll also get great results using index devices to cut multiple slots on a tablesaw or crosscut saw, or when you're routing evenly spaced flutes or reeds around a turned chair or table leg.

▶ MICRO-ADJUSTERS

Ever need to tweak a fence or stop block to an exact position—say, to set the distance between a fence and a part to set an exact measurement? Rather than just tapping the fence one way or another, use a micro-adjuster to move it very precisely over a short distance, so you can fine-tune the setting before locking the fence in place. The micro-adjuster shown here is designed to fine-adjust the fence on a router table, drill-press table, or other machine table. It can be adapted to work with many different jigs and machine setups.

One half of the adjuster is a block that screws to the back of the fence. A hole has been drilled through the block, intersecting the

edge, as shown, to create a notch. The other half of the adjuster is a U-shaped clamp made from plywood scraps, which mounts to the edge of the jig or machine table, locking in place with a carriage bolt and hand knob. A pointy-ended, goose-necked lever of thin tem-

Attached to the edge of a drill-press jig table, a micro-adjuster allows precise positioning of the fence.

pered hardboard attaches atop the clamp, pivoting on a wood screw.

To use the adjuster, roughly position the fence as desired and then mount the clamp block so that the lever engages the notch in the fence block. Rotating the lever moves the fence in very small increments, making it easy to get the exact fence setting you're after.

The U-shaped clamp portion of a micro-adjuster is glued up from three small pieces of plywood.

Flip Stops

Want to saw parts to length without marking them individually? Then fit the extension fence on your power miter saw or radial arm saw with a T-track, self-stick measuring tape, and flip stops. Start by screwing lengths of aluminum

➤ See *"Miter Saw Extension Tables"* on p. 127.

T-track (the kind that accepts the head of a ¼-in. bolt) to the top of the fence on either or both sides of the blade **(A)**. Make the front edge of the track flush with the fence face. Make the flip stop's sliding blocks by first cutting a 12-in. length of 1¼-in.-thick solid stock that you've ripped to the same width as the track. On the tablesaw, cut the bottom of this strip to form a ⅛-in.-high, ¼-in.-wide raised rib **(B)** that matches the position of the slot on the T-track. Now crosscut the strip into 2½-in.-long blocks, which will make four stops. Drill a ¼-in.-dia. hole through each block, centered on the bottom rib, then fit each block with a 2-in. x ¼-in. bolt and small hand knob. Make the L-shaped flip stop arms from ½-in. or ¾-in.-thick plywood **(C)**, sizing it to span the width of the T-track and the height of the fence. Saw a ⅛-in.-deep by ½-in.-wide notch in the shorter leg to accommodate a cursor plate. Drill a hole in each arm and use a washer-head screw to attach it to the sliding block. Mount the arm on the right side of the block for use to the left of the saw, and vice versa. Make each ½-in.-wide cursor plate from ⅛-in.-thick acrylic plastic, cutting it about 1¼ in. long (longer for miter flip stops; see the variation on the facing page). Use an awl and a small square to scratch a cursor line in the plate about an inch from one end and then highlight it with a colored fine-tip marker pen **(D)**. Drill a counter-sunk hole through the cursor plate and mount it

into the notch on the underside of the flip stop using a small flat-head screw. With the sliding stop locked on the track at an exact distance relative to the blade (say 17 in.), attach the stick-on tape measure to the top of the T track so that the mark for that distance (17 in.) is directly under the cursor **(E)**. Now the flip stop can be set for accurate crosscuts anywhere along the track. By flipping the stop up out of the way **(F)**, you can trim stock ends or cut parts to longer lengths while saving your stop settings for reuse later.

E

F

[VARIATION] **Some parts are too short to cut to length using a regular flip stop on an extension fence track. In those cases, you can screw and glue an extension bar and stop face to a flip stop arm (without a cursor) (A). The bar should be long enough to place the flip stop arm in front of the extension fence; stick the part of the tape measure left over from the regular flip stop track (above) and apply it to the face of the fence. The tape should be positioned so that the edge of the stop's arm shows the correct measurement. When cutting mitered frame members, it's often desirable to size frames by their inside dimensions. To create a stop for this, mount a triangular block to the end of a flip stop arm (B), and fit the arm with a longer cursor that shows the inside length of the frame member on the tape measure.**

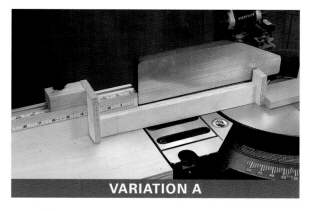

VARIATION A

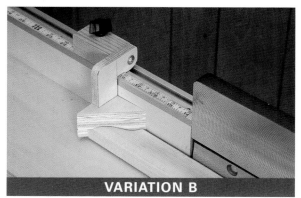

VARIATION B

Extension Stops

You can increase a fence's capacity for holding stop blocks by adding aluminum extension rails and sliding stops. One good example of this is the sliding miter jig.

▶ See *"Miter Sled"* on p. 93.

To extend its capacity using a T-track extrusion, start by cutting out a rectangular mounting plate from a scrap of ½-in. or ¾-in. plywood. Make the plate as long as the back edge of the jig's existing fence but make it ¾ in. to 1 in. wider than the fence is high. Drill a pair of widely spaced ¼-in.-dia. holes near the top edge of the plate for mounting T-track extensions. Attach the plate to the back of the fence with a pair of screws (A). Depending on the extrusion you use, mount the T-track either with a couple of wood screws or a pair of hex-head bolts secured with a pair of small hand knobs. Either make a sliding stop for the track, or use a commercial or shopmade flip stop to position long workpieces for accurate-length miter cuts (B). To add an extension stop rail to your tablesaw's standard miter gauge, cut a 2½-in.-wide x 18-in.-long mounting plate from a piece of ½-in.-thick plywood or MDF. Drill a pair of holes in the plate (for mounting an extruded aluminum track) and then screw the plate to the face of the miter gauge (C). Attach a 2-ft. to 4-ft. length of Wide T-Track to the plate, securing it with two-¼ in.-dia. bolts and hand knobs. For repeatable crosscuts, mount a stick-on measuring tape to the top of the track and set a flip stop at the desired location (D).

Eccentric Stops

An easy way to make an end stop with a small range of fine adjustment is to use the eccentric cam principal: Drilling an off-center hole in a round disc or dowel **(A)** produces a stop that will shift the position of a part. An eccentric dowel stop is a good choice for jig setups where you want to fine-tune the distance between the part and the cutter, or to readjust the stop to compensate for blades/cutters of different thicknesses. For example, eccentric dowel stops are used on the pivoting arms of the biscuit slotting table to adjust the position of slots cut on face frame stiles. (See p. 133.) Eccentric stops are also a useful addition on a sliding tenoning jig **(B)**, where they allow tweaking of the position of the shoulder cut. Another style of eccentric stop can be used to shift the position of a workpiece by selected intervals. This is useful for drilling several holes at precise distances from one another, or to shift a frame member between cuts when you're chopping a long mortise on a drill press fitted with a mortising chisel setup. To make a four-position mortising stop, measure and mark the desired dimensions on a piece of plywood, and then trim it to size **(C)**. Mount the stop to

▶ See *"Four-Sided Base"* on p. 161.

the mortising fence using a handscrew threaded into a T-nut **(D)**. Rotating the stop changes the location of the workpiece relative to the mortising chisel, with each edge of the stop determining a different distance.

[TIP] **The farther off-center you drill the mounting hole on an eccentric stop, the greater the range of adjustment.**

A

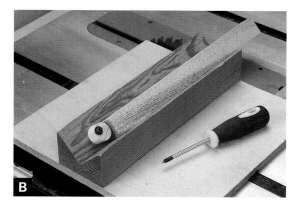

B

C

D

A

Fence Angle Stops

Adding a stop to a pivoting fence eases precise setting of the fence to commonly used angles. Start by positioning the fence at the desired angle. Take a test cut, checking it with a square or protractor and tweaking the fence position as necessary. Finally, drill through the fence into the jig's base, and then ream the hole with a tapered reamer **(A)**. Now you can slip a tapered pin into the hole to lock the fence at that angle.

▶ See *"Taper Pins"* on p. 26.

To set repeatable distances relative to a bit, you can incorporate a stop into a pivoting fence system like the one used on the quickie router table on p. 135.

▶ See *"Quickie Router Table"* on p. 135.

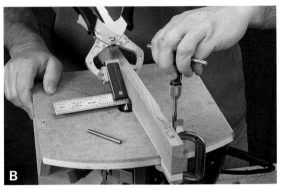

B

In this system, the fence registers against a stop block, which can be pinned to any one of a series of tapered holes drilled into the jig table. First, lock the fence in place at a precise distance relative to the center point of the bit. Drill a hole in a hardwood block, and clamp it to the table behind the fence **(B)**. Using the hole as a guide, drill through the jig table and then ream the hole to accept a taper pin. Repeat as desired to set the fence at other common distances (¼ in., ½ in., etc.) **(C)**. To easily cut a series of evenly-spaced slots or dadoes in a panel, first cut a strip of MDF or plywood as wide as the desired cut spacing and then crosscut the strip into a series of blocks. Set the tablesaw's rip fence for your first cut, and line up the blocks to the right of it, clamping the last block to the saw table **(D)**. After each cut, remove a block and reposition the fence against the next block to take the next cut.

C

D

Depth Stops

Wooden sleeves will limit the cutting depth of drill bits. Make them from dowels bored concentrically to create a hollow cylinder that slips over the bit **(A)**. To make a fixture to hold the dowel, drill a hole into a 2x4 near one end. Then saw a slit inward from the end to intersect the hole, and clamp the jig to the drill-press table. Insert a section of dowel, and secure it with a C-clamp. Using the bit you're making a sleeve for, bore through the dowel. Crosscut the hollowed dowel to create sleeves of various lengths. Using a single sleeve, you can fine-tune your boring depth by the amount of the bit shaft that you insert into the drill chuck. A sleeve-like stop collar can be used to regulate the depth of a shaping operation, as when dowel pointing.

➤ See *"Dowel-Pointing Guide"* on p. 89.

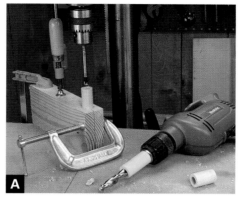

Drill a hole the size of the dowel through a small block of wood, saw a slit in from the edge, and add a screw to clamp the collar to the dowel **(B)**. A handy way to control boring depth on the drill press is with stops that slip over the machine's depth rod **(C)**. This allows you to reset the travel of the drill press quill (and hence, boring depth) much more quickly than if you had to reset the stop nut on the threaded rod. Make a set of these stops from a 1-in.-square strip of wood. Cut a channel (wide and deep enough to fit your machine's rod) down the center of the strip **(D)**. Now crosscut the strip to make stops of various common lengths (½ in., 1 in., 1½ in. etc.). Apply sticky-backed magnetic tape inside the channel to make it stick to the steel depth rod.

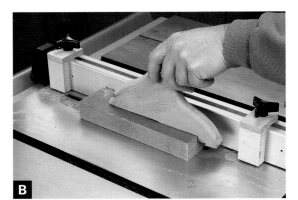

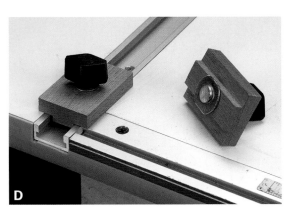

Travel Stops

Stops are an important jig device for limiting the length of cuts made on the tablesaw, bandsaw, shaper, router table, or even the jointer. The simplest approach is to clamp a block of wood to a machine's fence to serve as a stop that limits the length of rabbets, grooves, and shaped cuts. However, if your fence has a T-track, it's worth fitting it with a pair of track-mounted adjustable stops. Make the L-shaped stops from plywood, drilling out the top portion for a T-bolt, or for a threaded handscrew and T-nut that fits the track on your fence **(A)**. Use a single stop to terminate a cut shy of the end of the workpiece. To stop the cut short of both ends, use a stop at each end of the fence **(B)**. (For long workpieces, you'll likely have to mount a longer T-track atop your fence.)

It is both useful and safer to limit the travel of a miter gauge or jig sliding in a machine table's miter slot, say, when stopping a dado short of the edge of a bookcase side. If your machine has a T-style miter slot, then you can make a travel stop from a 1½-in. wide block of wood with a counter-bored ¼-in. hole in the center. On the tablesaw, cut a ¾-in. wide, ⅛-in. high tongue on the bottom surface **(C)**.

Next, file or grind two parallel edges on a ¼-in. fender washer so that it'll fit the T-slot. Hold the washer on a ¼-in. bolt with a nut. Grind the head of the bolt down until the assembly will slide in the T-slot, and mount it to the stop block and to a hand knob. Slide the washer into the T-slot, position the stop, and tighten the hand knob to lock the stop in place **(D)**. You can also make stops for non-T-slot miter grooves: Cut a short strip of hardwood that fits snugly in your machine's groove (most are ¾ in. wide), and then drill a ⁷⁄₁₆-in.-dia. hole through the middle. On the bandsaw, cut down the middle of the strip and ½ in. past the hole, and then drill a small hole at the end of the cut (to prevent the stop from splitting). Using a round file, elongate the ends of the big hole until it's oval-shaped **(E)**. The lock for the stop is made by cutting a ½-in.-dia. dowel to 1¼ in. long and then cross-drilling it for a 3-in.-long, ⁵⁄₁₆-in. dowel handle. File the bottom end of the dowel flat on two parallel sides so it'll fit into the oval hole. To lock the stop, simply turn the handle until the strip expands against the sides of the miter slot **(F)**. A magnetic stop is an alternative way of stopping the travel of a sliding jig or workpiece on tools with steel or cast-iron machine tables. Cut a base for the stop from ¾-in.-thick plywood and bore four shallow

> ▶ See *"Rare Earth Magnets"* on p. 32.

¾-in.-dia. holes to fit disc-style rare-earth magnets. Glue the magnets into the holes with epoxy or thick cyanoacrylate glue, leaving them slightly proud of the surface **(G)**. Attach a handle to the stop for easier positioning and removal. (Those magnets really stick to the table!) **(H)**.

> ⚠ **WARNING Always use a stop to limit the travel of a sliding jig to prevent it from sliding past the machine's blade or cutter and exposing you to danger.**

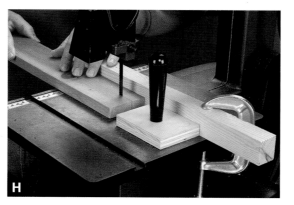

Routing Stops

There are lots of reasons to stop the forward motion of a router during a cut—to limit the length of a groove, to stop a rabbet cut short of the end of the workpiece, or to prevent shaping the part of an edge that must mate with another part. The basic router stop shown **(A)** clamps to the edge of the workpiece and stops the cut when the router base contacts it. Install a T-nut in the bottom and rout a slotted hole through the top of each 2-in. x 4-in. piece of ½-in.-thick plywood to accept the threaded handscrews that clamp them to the workpiece. Cut stepped blocks on the tablesaw **(B)** and glue these to the slotted half of the stop. This makes the stop adjustable for clamping onto ½-in.-, ¾-in.- or 1-in.-thick stock.

Another useful router stop setup uses a length of T-track and two sliding stops **(C)**. Make the two stops from 2-in. x 6-in. pieces of ½-in.-thick plywood, each drilled with a ¼-in.-dia. hole to accept a hex-head bolt and hand knob that locks it to the track. Glue on a spacer (cut as thick as the T-track) to shim up the underside of each stop where it extends over the work. To use this router-stop jig, clamp the track to the top of the work, and adjust the distance between stops to set the length of the router's cut.

You can regulate the length of slots, rabbets, and profiles cut into the edges of workpieces on the router table using a pair of stops mounted on the fence (as shown in photo B on p. 198). But a cam-stop setup, which eases the cutter into the workpiece in a slow and controlled way, yields cleaner stopped cuts for edge profiles or joinery, using large bits on a router table or shaper. This jig consists of two parts: an auxiliary fence with a beveled stop pin mounted above the bit, and a carriage that clamps atop the workpiece to regulate the cut **(D)**. Make the fence from a piece of MDF or particleboard cut to fit over your router table's existing fence. Using 1½-in.-thick stock, cut a 1-in.-wide pin with each of its ends beveled at 45 degrees. Glue and screw the pin to the auxiliary fence, centering it above the bit. Make the carriage from 1½-in.-thick stock cut a little longer than the workpiece. Cut a 1-in.-deep recess into the fence-facing edge of the carriage, beveling the ends at 45 degrees. The length of the recess determines the length of the routed portion of the work: As the stop pin contacts the non-recessed ends of the carriage, it forces the work away from the bit, thus starting and stopping the cut. To make the workpiece easier to mount to the carriage (for routing multiple parts quickly), add a small stop strip to one end and screw on a small quick-release toggle clamp that presses the work against the stop. Adding a stop strip to the carriage at the back edge makes the workpiece easier to align in the jig. With the carriage firmly clamped to the workpiece, start the cut by pressing the non-recessed edge of the carriage against the fence pin. Push the carriage forward, and the cut starts as the beveled part of the recess slides down the beveled pin **(E)**. Maintain pressure against the fence as the cut progresses **(F)**, until the beveled sections of recess and pin ease the work away from the bit at the end of the cut.

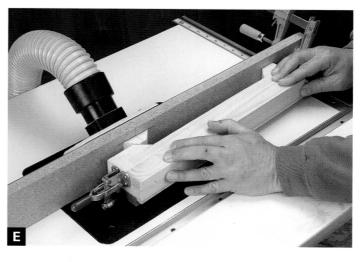

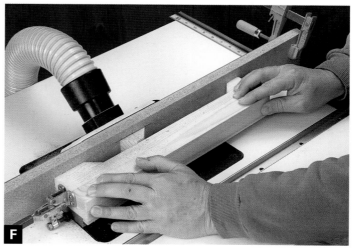

Hole-Spacing Jigs

It's easy to drill a series of evenly spaced holes for pegs, adjustable shelf pins, or decoration using a hole-spacing jig on the drill press. The simplest kind of spacing jig uses a spring-loaded pin, made from a dowel with a tapered end. The pin is glued to one end of a springy wood strip, while the opposite end is screwed to a fence that is clamped to the drill-press table. After the first hole is drilled, the workpiece is moved so that the pin engages the hole, and then a new hole is drilled (**A**). The process is repeated as many times as desired. The distance between the pin and drill bit determines the spacing of the holes. The only disadvantage to this approach is that the springy strip can create cumulative error if it flexes in use.

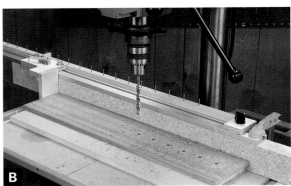

A more complicated but more accurate hole-spacing jig uses a nail strip and a sliding stop to reference the workpiece relative to the drill bit (**B**). Make the nail strip by cutting a 1½-in.-wide piece of ¾-in.-thick solid stock long enough to accommodate the workpiece. Drive a series of finish nails into the top edge, spaced at the desired hole spacing (**C**). (Drill small pilot holes first, so the nails drive straight.) Attach a pair of cleats to the nail strip, and mount the strip to a T-track attached to a drill-press fence. Next, make the sliding stop from ½-in.-thick plywood: The sliding stop consists of a stop block that is glued to an L-shaped top piece (**D**). The top piece attaches to the T-track with a couple of ¼-in. T-nuts and hex-head bolts that allow it to slide side to side. A pivoting arm assembly—spring-loaded with a rubber band—serves to register the stop against a nail (**E**). To use the jig, first place the workpiece against the sliding stop and drill the first hole. Then slide the stop to the right just past the next nail, slide it back so that the pivoting stop is flat against the nail, and drill another hole. Repeat until all holes are drilled.

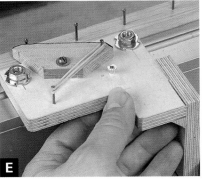

Kerf Spacing

A basic fence fitted with an indexing pin is great for spacing narrow grooves cut in drawer sides for dividers or cutting dadoes for thin spice rack shelves. To make a groove-spacing jig for your tablesaw's miter gauge, first make a 3-in.-high x 24-in.-long auxiliary fence from ¾-in. or 1-in.-thick straight stock. For easy adjustability, screw a short length of T-track to the rear of the fence, and mount the bar to the miter gauge head using a pair of ¼-in.-dia. bolts. Place the miter gauge into your tablesaw slot and crosscut a series of ⅜-in. deep grooves spaced exactly 1 in. apart along the length of the bar **(A)**.

> See *"Box Joints"* on p. 101.

I spaced the cuts using fence stop blocks to position the end of the bar relative to the blade, in a manner similar to that described on p. 196. Fit and glue a small hardwood pin snugly in the desired slot, fine-tune its distance relative to the blade, and you're ready to slot a workpiece. As you cut, set each previously cut kerf over the pin to make the next cut **(B)**.

You can use a variation of this jig on the radial arm saw or sliding compound miter saw, to cut closely spaced kerfs into a strip or panel in preparation for bending it **(C)**. First, cut a slot into a wooden auxiliary fence that you've sized to fit over your saw's regular fence. Fit a hardwood pin into the slot and clamp the fence to your saw, positioning the pin as far from the blade as the desired kerf spacing **(D)**. Then make the cuts as previously described, slipping the pin into each previously made kerf.

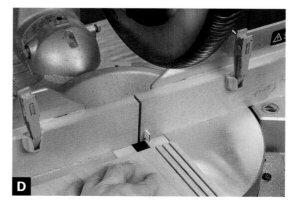

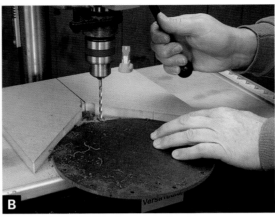

Lathe-Indexing Plate

You can use a routing jig to cut flutes or reeds into a spindle-turned leg or post mounted on the lathe. However, you need to be able to lock the lathe's headstock rotation at specific intervals to space the cuts evenly around the workpiece. If your lathe doesn't have a built-in indexing plate, you can make one. Start by cutting a large disc from ¼-in.-thick tempered hardboard, using a circle-cutting jig.

> See *"Circle Jigs"* on p. 116.

Set a pair of dividers to half the radius of the disc and mark a series of 12 evenly spaced lines around the disc's circumference **(A)**. Use a disc fence to drill ³⁄₁₆-in.-dia. holes about ¼ in. in from the edge of the disc, locating one hole at each line **(B)**. Mount this disc to a faceplate **(C)** between the lathe's headstock and the spindle you're planning to turn.

> See *"Disc Fence"* on p. 83.

To make the stop bar assembly that locks the index plate **(D)** (on a twin-rail, flat-bed lathe), cut a vertical bar that's an inch longer than the distance from your lathe's bed to the spindle center. Cut a centered ⁹⁄₃₂-in.-wide groove into the top end of the bar (for the plate), and screw the other end to a cross bar that spans the lathe bed and positions the slot over the plate. Mark and drill a hole through the top end of the slotted bar that's exactly level with the lathe spindle, and then position the stop so it engages the plate and ream all 12 holes for a tapered pin. (See *"Taper Pins"* on p. 26.) After turning your spindle, position and lock it in place with the pin, choosing whichever series of holes in the plate that yields the desired spacing **(E).** Use the router fluting jig to guide the router cuts.

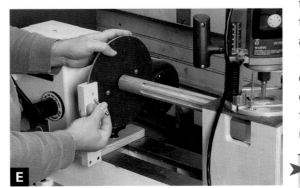

> See *"Fluting Jig"* on p. 157.

Clamping and Alignment

BEFORE YOU CAN ACCURATELY cut, shape, bore, or sand a workpiece, you need to hold it firmly in place. That not only means clamping it firmly to a jig, fixture, or machine table, but positioning it precisely, relative to its cutter, blade, bit, or abrasive belt or disc. While fences and stops do this in a general way, positioning blocks are more specialized and are made to suit specific workpieces or machining situations.

Such blocks help to align parts so that holes are bored in the right places or cuts are accurately made, as shown in the bottom photo. Because all woodworking operations —both hand and machine—generate some force, workpieces must almost always be restrained from moving due to clamping force, as discussed in a moment. For light operations, such as boring small holes, sanding edges, or crosscutting thin strips, hand pressure may be all that's needed to secure the part.

Besides commercial clamps, there are many clamping devices you can make that supply the holding force required by heavier machining tasks. Shopmade clamps are particularly great, because you can incorporate them directly into your jigs and fixtures. Wood wedges, cleats, and cam clamps (see the top photo on p. 206) are easy to make and can produce an impressive amount of force to hold parts tightly, assuring a more accurate and safe machining operation. For

Alignment Devices

➤ Positioning Blocks (p. 207)
➤ Assembly Blocks (p. 208)

Simple Clamps

➤ Handheld Parts (p. 209)
➤ Handscrew Clamps (p. 211)
➤ Through Bolting (p. 212)
➤ Wedges and Cleats (p. 213)
➤ Cam Clamps (p. 215)

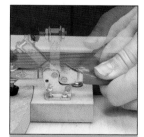

Production Clamps

➤ Toggle Clamps (p. 216)
➤ Air-Clamping Systems (p. 217)
➤ Vacuum-Clamp Systems (p. 218)

Used with a tablesaw's miter gauge, a V-block is handy for positioning square stock diagonally for tenoning.

the more demanding clamping needs of production woodworkers, the photo-essays at the end of this section explore the use of quick-action toggle clamps (see the bottom photo on p. 206) as well as air and vacuum clamping systems.

Shaped like a circle with a pointy handle sticking out of it, eccentric cam clamps are cut out of plywood on the bandsaw.

To increase the holding efficiency of production toggle clamps, apply anti-slip stair tread tape to the surface beneath the clamping head.

Restraining Workpieces

When you consider all the possible ways a workpiece can move out of position during a machining process, clamping it firmly can seem like a daunting challenge. Ultimately, a workpiece must be kept from moving in 12 possible directions, as shown in the drawing below. Machinists call this concept the "12 degrees of freedom." Fortunately, you don't need a clamp for each possible type of motion: Judicious placement of clamps and positioning devices will quickly limit the movement of the part. The goal is to use the minimum number of clamps needed to get the job done to keep clamping and unclamping from becoming a tedious task.

One way to reduce the overall number of clamps needed is to employ positioning devices, such as blocks, fences, stops, etc., where they'll support the work against the forces generated by machining. This way, the rotation of the bit, cutter, or blade actually presses the stock more tightly to the positioning device, eliminating the need to place clamps or fingers too close to the action.

Another way to minimize clamping is to use the forces generated by a cutter, blade, or bit during machining to limit some degree of motion of the workpiece. For example, the feed force and direction of a spinning drill bit press the work down against the bench or drill-press table, and its clockwise rotation tends to rotate the work, thus pressing one end of it more tightly to any stop or fence the part is bearing against. The only clamping needed in this case should keep the stock from lifting and the opposite end of the stock from rotating away from the fence/stop.

THE 12 DEGREES OF FREEDOM

When a part is held in a fixture, movement must be controlled in 12 different directions.

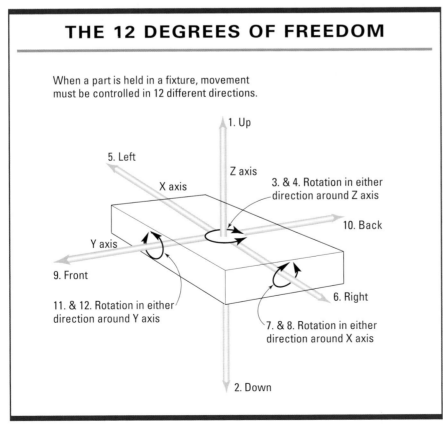

1. Up
5. Left
X axis
Z axis
3. & 4. Rotation in either direction around Z axis
10. Back
Y axis
9. Front
11. & 12. Rotation in either direction around Y axis
6. Right
7. & 8. Rotation in either direction around X axis
2. Down

Positioning Blocks

Before you can clamp a part down into a jig or atop a machine table, you need to align it relative to the blade, bit, or cutter. V-blocks are useful for positioning stock in several jigs, but they're also useful on the drill press to hold dowels steady for boring **(A)**.

➤ See *"Dowel-Pointing Guide"* on p. 89.

Another useful drill press device uses two ½-in.-dia. x 2-in.-long dowels as positioning pins, to automatically center boards under the bit **(B)**. Drill a pair of holes 6 in. to 12 in. apart in a plywood base to mount the dowels. Clamp the jig to the drill-press table with the bit in line with and exactly equidistant from the pins. To center a hole, set the work between the pins and rotate it until it's in firm contact with both pins.

Positioning blocks are also useful when you need to accurately locate a shaped workpiece such as a curvaceous table leg precisely in a jig for a machining operation. Two or more blocks fit around the ends of the part or against its edges will keep it from moving as it's cut. Cut the blocks out using a jigsaw or bandsaw **(C)**, and after carefully positioning the work on the jig, screw the positioning blocks down around the part **(D)**. It's best to locate one of the blocks near the part of the workpiece being cut, to keep the force of the blade from pushing it out of alignment. In many cases, the blocks secure the part well enough that hand pressure is all that's needed to hold it down during cutting. If more clamping pressure is needed, mount toggle clamps to press the part firmly down. (See photo C on p. 216.)

A

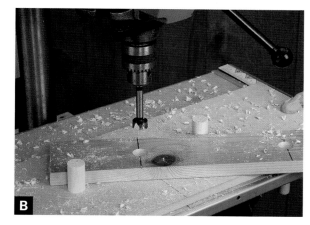

B

C

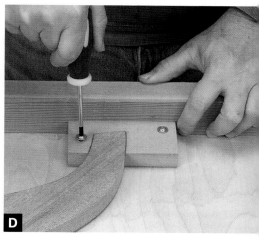

D

Assembly Blocks

Positioning blocks aren't good only for precisely aligning parts during machining operations. Keeping surfaces perpendicular and edges flush is imperative for successfully gluing and fastening several parts into an assembly. You can quickly fashion alignment blocks that will position parts accurately for any assembly situation and keep them in place until nails or screws are installed and/or clamps applied. In this example, two blocks are used with three parts joined together into a subassembly for a small cabinet (A). The blocks are cut from scraps of MDF or plywood, each grooved using a dado blade set on the tablesaw (B). The grooves, each cut slightly wider than the thickness of the workpiece, are aligned and oriented to hold parts square and flush (a little wax rubbed in each groove keeps glue from sticking). A small block glued and nailed at the bottom edge of each assembly block keeps the block and the part it's holding standing up straight and square to the assembly table.

Assembly blocks are also a godsend when you're gluing and fastening the mitered corners of a box, chest, drawer, or frame. The blocks not only hold the parts at exactly 90 degrees to one another, but they keep the miter tips aligned during clamping and fastening (C). Make a set of these blocks by cutting two perpendicular grooves in each block near the edges. For a narrow frame, make four blocks; for a wide-sided box or chest, make eight (for the top and bottom of each corner). Cut off the corner of each block where the grooves meet at a 45-degree angle (D). This prevents glued parts from adhering to the blocks, and provides better clearance for driving nails or screws near the edges of the parts.

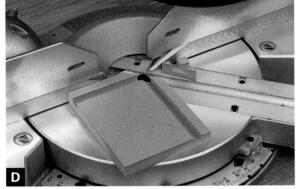

Handheld Parts

The human hand is an amazing device. It can securely hold parts of all shapes and most sizes, even parts oddly angled or too irregular for conventional clamping devices. Handholding parts even saves us the time it would take to apply and remove clamps. The problem comes when trying to hold small parts: How do you to grasp the part firmly yet not get precious digits too close to flesh-hungry blades and cutters? A simple way of preventing parts from slipping out of position as they're handheld during cutting is to add a piece of PSA sandpaper or anti-slip stair tread tape **(A)** to the surface of the jig base, fence, positioning block, or stop they're held against.

Unless you're using a special jig designed to handle small parts, you can make it much safer to cut small blocks and strips to length by making a part-holding stick. Cut a strip of stout hardwood and cover one surface of each end with PSA sandpaper **(B)**. Cut a small shim the same thickness as the part and use the part-holding stick to firmly press both the shim and part against the saw's fence **(C)**. If necessary, clamp a strip of MDF or plywood to the saw fence to support the part right next to the blade. This stick method works well, both with crosscutting saws and when you're using a miter gauge on the tablesaw.

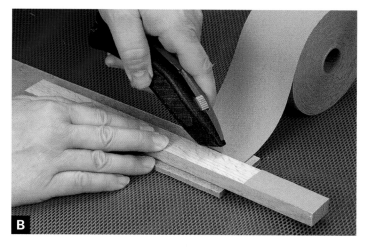

When boring on the drill press, you can keep a part steady by preventing it from rotating and from lifting when the bit is withdrawn. To keep it from rotating without clamping it, butt one or more edges of the part against a fence. Carefully hold the part down to the drill-press

(Text continued on p. 210.)

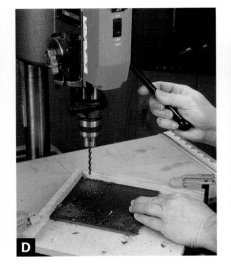

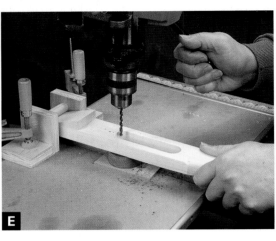

table with your hand positioned well away from the bit. When you're drilling multiple parts, using a fence has the added benefit of positioning the stock relative to the bit. By using a two-sided fence **(D)**, you can position a square or rectangular part so that holes are spaced accurately in all four corners.

As the teachings of Archimedes reveal, you can greatly enhance your ability to handhold smaller parts steady by employing a lever. This drill-press hold-down jig **(E)** is just the ticket for holding small parts rock-solid while boring them. The lever arm, made from a 2-in. x 15-in. piece of 4/4 hardwood, has a wedge-shaped end made by laminating a 3½-in.-long piece on one end and then cutting out a wedge profile. Make the slotted hole in the middle of the lever by boring two 1-in.-dia. holes 3 in. apart and sawing between them with a jigsaw **(F)**. For a more comfortable grip, cut out and round over the handle portion of the arm. Add some PSA sandpaper at the edges of the slot on the underside of the arm to further aid its clamping efficiency.

The arm levers against a base that's clamped or screwed to the drill-press table. Make the base by joining two 2¾-in.-high L-shaped supports, gluing them together from ¾-in.-thick plywood scraps with a 3¾-in.-long piece of ⅝-in.-dia. dowel **(G)**. Sliding the wedged end of the arm a little or a lot under the dowel adjusts its height to suit the thickness of the workpiece.

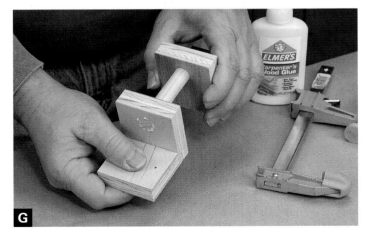

> ⚠ **WARNING** Take extra care when boring parts with sharp corners on the drill press. If the bit grabs and spins it, it could smash or cut your knuckles as well as ruining the bit.

Handscrew Clamps

Traditional wooden handscrew clamps come in a wide range of sizes and are great for clamping stock on a workbench or for securing a fixture to a machine table. In addition, they're actually great holding devices by themselves. A handscrew is terrific for firmly clutching strips, blocks, or discs too small or awkward to hold by hand. The surface of the clamp even keeps flat-bottomed parts perpendicular for hole boring **(A)**. Gripping the large body of the clamp keeps your hands well away from the bit. Since its jaws adjust independently, you can use a handscrew to hold tapered or irregular parts, for example, to rout grooves or shape edges on a router table **(B)**. Because the clamp is made of wood, there's no serious harm if the router bit accidentally comes in contact with the clamp jaws.

You can also use wooden handscrews to hold small or irregular parts firmly in a jig, fixture, or on your benchtop. To mount the clamp to a surface, bore an oversized hole (or rout a slot) perpendicularly through each of the clamp's jaws. Fit bolts or drive long wood screws through these holes **(C)**. To make a handy holding fixture, cut a 10-in. x 12-in. baseplate from ¾-in.-thick MDF, and set a length of T-track into a groove so it's flush with the surface.

▶ See *"Extruded T-Tracks"* on p. 29.

Secure the handscrew to the track with a pair of T-bolts and hand knobs, and clamp the base into your bench vise. The clamp firmly holds small parts for carving, planing, sanding, etc. **(D)**, while the T-track lets you reposition the clamp as required.

[TIP] A pair of locking pliers is great for securely holding tiny parts, screws, etc., for sanding, filing or grinding.

Through Bolting

Sometimes it's impractical to use any kind of clamps to hold a part firmly in a jig, because the part is either too irregular to hold securely, or the clamps get in the way during machining. A practical solution is to bolt or screw the part directly to the jig or fixture.

If the part has holes in it, these can serve as temporary mounting sites. If it doesn't, you can cut the part to include mounting sites. On rectangular parts and strips, simply cut the part longer than final length, and bolt it near the ends. For irregularly shaped parts and complex jig setups, include two or more rectangular lugs in the outline of the part to bolt through **(A)**.

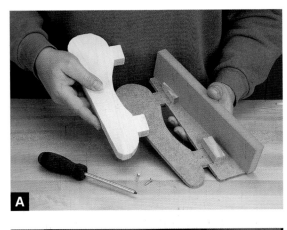

The jig shown here holds and orients an irregular part for two different operations. For the first operation, the part is screwed directly on top of a base, using wood screws driven from underneath. (A pair of positioning stops help orient the part in the jig.) The base serves as a template when you're trimming the outside edge of the part to final size on a router table, using a bottom-piloted flush-trim bit **(B)**. Once that's done, the jig is turned 90 degrees to rest on an angled strip glued to the back edge of the template. (This strip also serves as a handle for controlling the part during routing.) The strip is angled relative to the part so that it will position the part's tapered top edge level, so that several holes can be bored at 90 degrees into this edge on the drill press **(C)**. Once the machining is done, the part is unscrewed from the jig, the lugs are cut off **(D)**, and the part is finish-sanded.

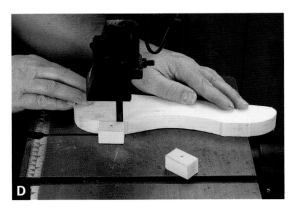

Wedges and Cleats

You can exploit the ancient principle of using wedges individually or in pairs to create a clamping force that's useful for securing parts into all manner of jigs and fixtures. Basic wedges of any size needed are easy to cut from any rectangular scrap of wood. Draw a line diagonally between two opposing corners of the block, and cut out two identical wedges on the bandsaw **(A)**. The longer and skinnier the block, the farther you have to drive the wedge to engage the clamping action, but the greater the clamping force and the less the likelihood that the wedge will come loose. For general use, cut wedges from blocks with about a 5-to-1 length-to-width ratio. To make a clamp from a single wedge, drive it against any part set between two stops. Use a mallet to drive the wedge either against the part itself or, in the case of curved parts, against a positioning block cut to match the part's profile **(B)**. This way, you can use dowels set into a carrier board to hold an irregular part as the jig passes through a thickness planer or wide-belt sander.

Driving a pair of identical wedges together also creates a useful clamping force. For example, you can use two pairs of wedges to press a series of slats firmly against one another in a frame **(C)** in preparation for gluing on a canvas backing to create a tambour door. As each wedge pair is driven, it expands in width while the outer surfaces remain parallel, thus applying flat, even pressure to the edge of the slats.

You can also employ a pair of wedges to hold a part flat to the surface of a jig, either horizontally on a base **(D)** or vertically to a fence. Each clamping device has a fixed and a loose wedge. To make the fixed one, cut an angled notch out of a hardwood block by first cutting it partway through

(Text continued on p. 214.)

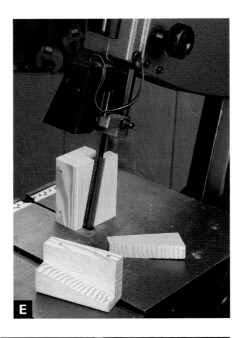

E

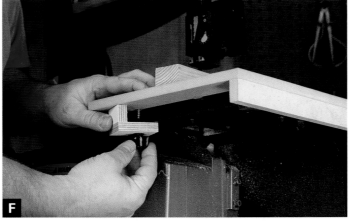

F

G

lengthwise on the tablesaw, and then sawing out the angled portion on a bandsaw with its table tilted **(E)**. Cut another wedge that matches the angle of the notch. Screw the notched block to the jig, positioning it over the edge of the part. (You can also use the block as a stop to hold the part in the desired position on the jig, thus serving double duty.) To clamp the part down, drive the loose wedge into the notch. To release parts from this or any wedge-clamping setup, simply drive or pull them apart.

Like wedges, cleats are simple devices useful for securing a part into a fixture, as well as for attaching a jig to a machine. You can make a basic cleat from two small scraps of plywood glued and screwed into an "L" shape. Cut the short leg of the L as high as the thickness of the workpiece, and drive a single screw through the other leg to clamp the part. Basic cleats like these are very handy for attaching an extension table or jig table to a machine table **(F)**. Basic cleats also work well with T-track. Just install a length of track in the base or other surface of a fixture or jig, and use T-bolts to secure the cleats in place **(G)**. The track lets you attach as many cleats as needed and position them wherever you wish along length of the workpiece.

[TIP] To make clamping wedges easier to drive apart, bore a hole into the side of each wedge, and put the tip of a screwdriver into the hole and tap it with a mallet.

Cam Clamps

You can make a simple yet versatile clamp that works on the eccentric cam principle by sawing a disc with a handle from ¾-in.-thick plywood. Drill an eccentric hole, and mount the clamp to a base or other jig surface with a bolt or screw **(A)**. The farther off-center on the disc the hole is drilled, the greater the range of part thicknesses you can clamp with it. However, the closer the hole is to center, the greater the clamping force. Use this kind of eccentric cam clamp to hold a workpiece against a fence or stop. Simply turn the handle one way to clamp the part, and the other way to release it.

Another very useful cam clamp design uses a locking lever based on the traditional luthier's wooden clamp **(B)**. Make the head for each clamp from a hardwood block that's at least 1 in. thick, 2 in. wide, and long enough to suit the width of the clamped part. (The head shown here is 7½ in. long.)

To make the clamp's jaw, first drill a small hole ¼ in. up from the bottom edge of the head and halfway down its length. Then, using the bandsaw, make a cut ¼ in. in and parallel with the edge until you intersect the hole. With a dado blade, cut a ⅜-in.-wide slot two-thirds down the length of the head (thickness-wise), for the clamping lever **(C)**. Set the blade height to leave ¼ in. uncut. Cut the cam lever from dense hardwood, shaped as shown **(D)**, and drill an eccentric hole through it for a nail that allows it to pivot in the head. Mount the clamp to your jig with a bolt set through a hole drilled in the other end of the head.

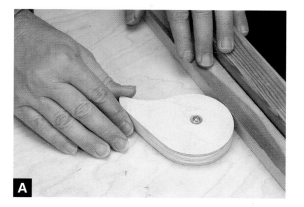

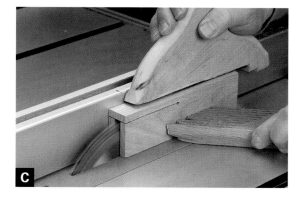

Toggle Clamps

Originally designed for production work, toggle clamps are extremely useful for holding workpieces in jigs as well as for clamping jigs to machines. Also called "quick-action clamps," toggle clamps use a multi-link mechanism to rapidly engage and disengage a rubber-tipped clamping head with the flick of a lever **(A)**.

Toggle clamps come in many styles and sizes **(B)**, each designed to suit a specific range of applications. Push-action clamps have a horizontal plunger that's useful for holding parts against a fence, stop, or other surface. Lever-action-, T-handle-, and post-style clamps all have clamping heads that swing down and press parts flat to a surface. The differences lie in the way the handle operates, and its position when locked or unlocked. Depending on the mounting location, you might choose a T-handle style over a lever-style model simply because there isn't space behind the clamp for the lever style's long locking handle.

Most toggle clamps mount to the surface of a jig or fixture with four screws **(C)**. You can also mount a toggle clamp to a T-track by screwing it to a block and then attaching the block to the track with a T-bolt **(D)**. When using a toggle clamp to secure a workpiece in a jig, positioning it close to the blade or cutter helps keep the part from moving during machining and also keeps your hands out of harm's way. Adjust the clamp's rubber-tipped head to apply enough pressure to hold the part firmly without making it difficult to operate the lever. You can improve the clamp's holding power by adding stair-tread tape or PSA sandpaper directly under the clamping head. (See p. 206.)

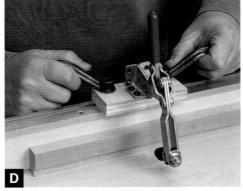

[TIP] To clamp thicker workpieces, add a spacer block between a toggle clamp and the jig.

Air-Clamping Systems

Usually reserved for production situations, air clamping systems are highly useful even for small-shop jig building. Air clamps have endless applications: holding parts firmly atop a workbench for routing, planing, or sanding **(A)**; clamping parts securely into a sliding tablesaw cross-cutting carriage, or holding frame members tightly against a drill press fence for boring or mortising operations **(B)**. Air clamps are also extremely useful for holding parts in assembly fixtures during glue-up or fastening operations.

Air clamps are basically toggle clamps that are fitted with a double-action, air-powered cylinder that locks and unlocks the clamp. The clamp closes or opens when air pressure—controlled by a four-way valve in a foot switch or joystick controller—is fed into one end of the cylinder or the other, as shown in the drawing at right. Since air is only consumed when the clamp opens or closes, air clamps can run on a very small compressor (½ hp or larger).

Air clamps come in various sizes, and in both lever-action and plunge-action styles, like the toggle clamps discussed in the previous photoessay. They also have various stroke lengths: Longer-stroke clamps can handle a wider range of part thicknesses and retract farther than short-stroke clamps, making it easier to load and unload workpieces. Many types stay locked even after the air supply is disconnected, which is great for glue-up fixtures left loaded overnight.

A basic air-clamping system requires a compressor, a filter/regulator, air clamps, a four-way valve controller, and fittings and tubing to connect them together. Fittings that work with ¼-in. nylon tubing are the easiest to use, since the tubing just plugs in, making connections very easy. When two or more clamps are used with a controller, you'll need to use T fittings to distribute the air to the various cylinder connections.

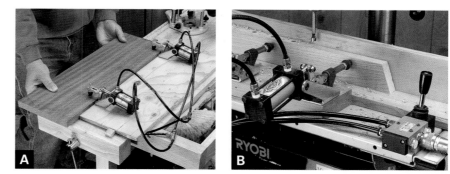

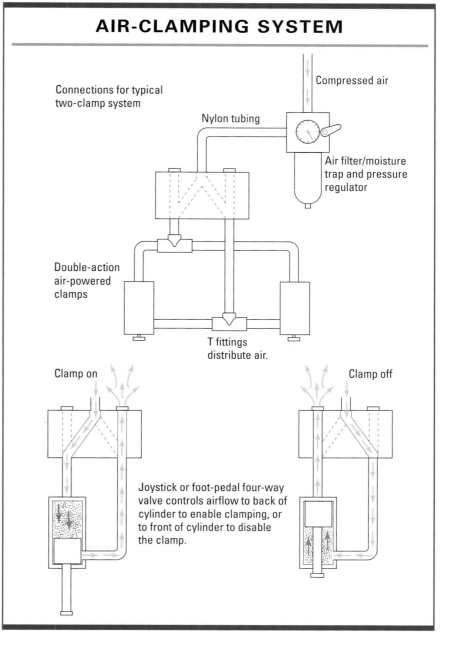

AIR-CLAMPING SYSTEM

Connections for typical two-clamp system

Compressed air

Nylon tubing

Air filter/moisture trap and pressure regulator

Double-action air-powered clamps

T fittings distribute air.

Clamp on

Clamp off

Joystick or foot-pedal four-way valve controls airflow to back of cylinder to enable clamping, or to front of cylinder to disable the clamp.

VACUUM-CLAMPING SYSTEM

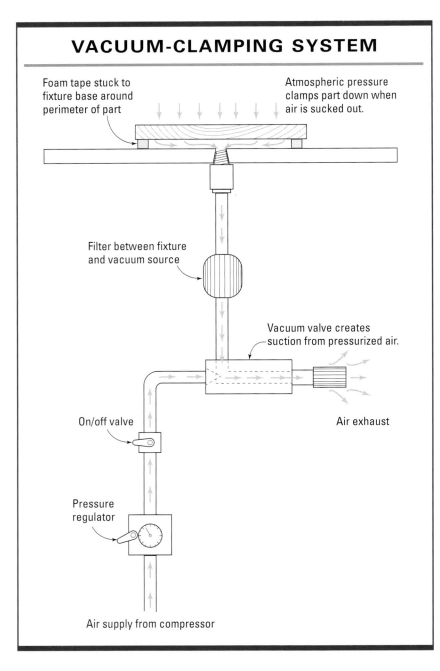

Foam tape stuck to fixture base around perimeter of part

Atmospheric pressure clamps part down when air is sucked out.

Filter between fixture and vacuum source

Vacuum valve creates suction from pressurized air.

On/off valve

Air exhaust

Pressure regulator

Air supply from compressor

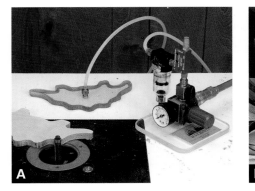

A

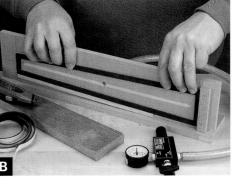

B

Vacuum-Clamp Systems

The pressure exerted by the earth's atmosphere—14 psi—is actually enough to hold parts securely during many woodworking operations. All it takes to harness this force is a vacuum. A vacuum-clamping system supplies and controls suction between a workpiece and whatever flat surface it's clamped to—a jig base, a template (A), a fixture, etc.

You can use a dedicated vacuum pump to generate the suction, but a more compact and economical alternative is a vacuum valve system as shown in the drawing at left. This compact device takes compressed air from a compressor and passes it through a regulator (set between 40 and 80 psi) and then through a small funnel-like Venturi that generates suction as the air blows through. A plastic hose connects from the valve's vacuum port to the clamping surface on the jig or fixture.

The clamping surface can be made from any non-porous material—plastic, hardboard, melamine, etc. A thin strip of foam rubber weather-stripping applied to the clamping surface around the perimeter of the workpiece forms a gasket that creates an airtight seal (B). The only condition is that the part must be at least 10 sq. in. in size, as atmospheric pressure won't hold smaller parts adequately.

To use this vacuum-clamping setup, turn on the compressed air and lightly press the workpiece down on the clamping surface until it's sucked flat. To prevent large, thin parts from bowing in the center from the pressure, add strips of weather-stripping to support the piece in the middle. Simply turn off the air supply to remove the part after machining.

[TIP] You can hold many small parts safely for sanding on a disc or belt sander by sucking them onto the end of a shop vacuum hose.

Fixtures

WHILE THE MAJORITY of this book focuses on jigs for cutting, drilling, shaping, planing, and sanding operations using stationary shop machines, portable power tools, and the occasional hand tool, this section will focus on fixtures. Although many of the jigs in earlier sections contain fixtures as part of their design, the fixtures in this section are primarily made for holding workpieces during machining and for aligning and supporting assemblies for glue-up. Some of them are actually forms for bending wood to shape.

A workbench wouldn't be much use without a sturdy vise or two bolted to it. But the end vise and side vise on a traditional bench don't always provide the best way of holding stock firmly during hand or power tool operations. Workbench fixtures are easy to make and provide a very useful way of securing a panel during routing, or for holding a narrow plank while you're planing its edge straight and true, as shown in the photo at right.

Most of these fixtures utilize shopmade or commercial clamping devices like those discussed in Section Eleven. These include cam clamps and toggle clamps as well as vacuum valve clamping systems. Using suction produced by compressed air flowing through a special vacuum valve is an especially useful technique employed in a couple of part-

Part Holding

➤ Workbench Fixtures (p. 221)
➤ Routing Fixtures (p. 222)
➤ Drilling Fixture (p. 224)

Assembly Fixtures

➤ Edge-Gluing Fixture (p. 226)
➤ Frame Clamping (p. 228)
➤ Coopered Panel Fixture (p. 229)

Bending and Lamination

➤ Bending Forms (p. 230)
➤ Lamination Forms (p. 231)

This benchtop fixture uses a fence, a wedged stop, and a pair of eccentric clamps to hold a board on edge for planing, shaping, routing, or sanding.

holding fixtures discussed in this section, such as the drill-press vacuum table shown in the top left photo on p. 220.

The assembly of parts—turning four mitered members into a picture frame, or

Relying on suction supplied by a compressed-air-powered valve, this handy vacuum table fixture holds panels and boards flat and secure during drilling operations.

This frame-gluing fixture uses cam clamps to apply pressure to the mitered parts.

This simple assembly fixture has concave struts that support beveled staves and holes for bar clamps that press the staves together as they're glued up into a curved door.

stiles and rails into a cabinet face frame—is another area of woodworking where fixtures are essential. Shop-built assembly fixtures are extremely useful for making common assembly tasks—such as edge-gluing boards to make a tabletop or panel—both quick and easy. They can even include built-in clamping devices, thus eliminating (or at least minimizing) the need for fumbling with loose clamps. For an example, take a look at the miter frame jig (photo C) on p. 228, which uses base-mounted eccentric cam clamps to press the corners of a frame together.

You can also build specialized fixtures for creating custom parts, such as for gluing up coopered, curved panel doors for a shapely vanity or dresser (see the photo, above left). Forms for bending wood and for gluing up bent laminations are also types of fixtures that are necessary for making curved or irregularly shaped parts from green wood, steamed wood, or glued-up veneers.

▶ GLUING GUIDE

If you've ever been frustrated while trying to put a neat bead of glue along the edge of a board in preparation for assembling a panel or cabinet side, then you'll really appreciate this simple gluing guide. Made to fit over the tip of a plastic bottle of yellow or

white carpenter's glue, the inverted-U-shaped guide slides along the edges of the stock, keeping the bottle's tip centered so you can concentrate on squeezing out just the right amount of glue. Make the guide by cutting a short strip of 4/4 hardwood about ⅜ in. wider than the stock you're gluing. With a dado blade, plow a centered slot down the strip that's a skosh wider than your stock width. Cut the strip into several 1-in. to 2-in.-long guides, and drill an angled hole in each to fit your glue bottle's tip.

Workbench Fixtures

You can't do good work if you can't hold your work steady on the workbench. Here are two basic workbench stock-holding fixtures. The first one holds boards and panels flat on the benchtop. The second supports long boards and strips on edge for jointing, sanding, routing, etc.

To make the first fixture, use a clean, straight-grained 2x4 that's no longer than your benchtop. Screw several toggle clamps to the flat side of the 2x4, spacing them evenly along its length **(A)**. Temporarily clamp the fixture down, or secure the ends of the 2x4 between your bench dogs. To clamp thinner workpieces, adjust the rubber tips of the toggle clamps, and/or use a piece of scrap stock under the workpiece as a shim **(B)**.

To build the on-edge fixture **(C)**, begin with ¾-in.-thick MDF or plywood, making the base 5 in. to 6 in. wide and 3 ft. to 4 ft. long. Glue and nail a 3-in.-wide fence strip 1 in. from one edge of the base. Reinforce the fence by adding some triangular glue blocks. Cut a 3-in. x 8-in. wedge from 2x4 stock, and glue and screw it to the base **(D)**, also reinforcing it with several triangular glue blocks. Glue up two layers of ¾-in.-thick plywood and cut out two or more 3½-in.-dia. x 5½-in.-long cam clamps.

▶ See *"Cam Clamps"* on p. 215.

Using long carriage bolts, mount these workpiece-holding clamps to the fixture's base **(E)** through counter-bored holes located 2½ in. from the fence. Clamp or screw the fixture to your benchtop, and press the end of the workpiece into the cleft between fence and wedge, and tighten the clamps. If they slip, add cloth tape to the edges of the cams.

A

B

C

D

E

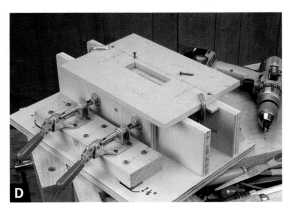

Routing Fixtures

Part of what makes routers such versatile work-shop power tools are the fixtures that both clamp and position workpieces during routing. A basic routing fixture works with many different kinds of shopmade templates for cutting mortises for joinery, as well as slots and recesses for installing inlays, mounting hardware, and more.

Make the jig's 12-in. x 18-in. base and two 4-in. x 18-in. fences from ¾-in.-thick plywood. Glue and screw the fixed fence to the base, locating it 2 in. from one edge and adding reinforcing glue blocks as shown **(A)**. The other "loose" fence clamps the workpiece using a pair of plunger-style toggle clamps like those discussed in Section Eleven. The clamps attach to a ¾-in. x 3½-in. x 8-in. plywood pressure plate, using a pair of bolts set through counter-bored holes in the plate, spaced 5 in. apart and driven into the end of each clamp's threaded plunger **(B)**. Screw the clamp bases to the top of a 12-in.-long piece of 2x4 **(C)**. By screwing the 2x4 to the fixture's base at various distances from the fixed fence, you can set the jig to clamp workpieces of different thicknesses. Attach the loose fence to the pressure plate with a couple of countersunk flathead screws.

▶ See *"Mortising Templates"* on p. 176.

Attach the routing template to the fixture's fixed fence with several screws **(D)**. To accurately position it relative to the fence face, temporarily clamp a stop strip to the template. Clamp the fixture to your workbench, slide the work under the template, and secure it with the two toggle clamps **(E).** (If the clamp handles don't clear the edge of the template, add a shim or two between the loose fence and pressure plate.) Raise narrower workpieces by sliding a spacer strip under them **(F)**. Leave a little space between the stock and the underside of the template so chips can escape during routing.

If you have lots of parts to rout, a vacuum hold-down fixture will save you a lot of time. Built atop a small bench or work stand, this fixture uses a vacuum valve powered by compressed air to suck a panel or other wide part down to its flat surface.

► See *"Vacuum-Clamp Systems"* on p 218.

Make the top from ¾-in.-thick melamine or MDF, sizing it at least as big as your largest work-piece. Attach an apron around the bottom edge of the top to keep it flat. Drill and tap a hole in the underside of the top to connect the vacuum hose **(G)**. Stick a band of ¼-in.-thick rubber or neoprene weather-stripping around the perimeter of the top (it must enclose at least 10 sq. inches for the vacuum to work).

Mount the top to a stand or worktable, fasten the vacuum valve to the base, and connect the vacuum hose. Control the suction with either an air pressure on/off valve or, for production work, a foot-controlled air valve **(H)** connected as shown in the drawing on p. 218. After turning on the vacuum, gently press the workpiece down to get it to seal against the weather-stripping.

The fixture is handy not only for routing parts but for planing, carving, or sanding them as well **(I)**. When you're done, turn the valve or press the foot pedal to shut off the vacuum and release the part.

> ⚠️ **WARNING Make sure the foot-pedal control for your vacuum table is set out of the way, so you don't accidentally step on it and release the part during powertool operations.**

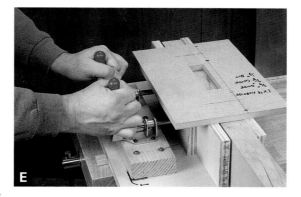

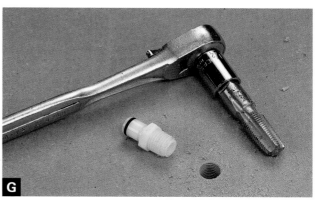

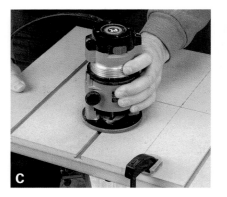

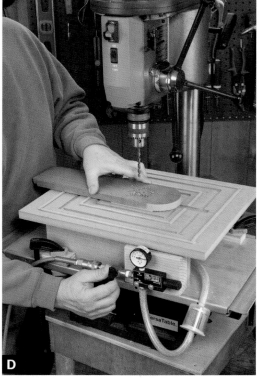

Drilling Fixture

Holding parts down securely on the drill press not only makes boring operations safer, but also assures more accurately placed, cleaner-edged holes. The drill-press fixture **(A)** uses T-bolts and hold-downs to keep parts flat and steady during boring.

Make the fixture's sturdy table by laminating together two layers of ¾-in.-thick MDF. To add the T-slots to the top, first plow at least four grooves across the table, using a straight bit and router or a dado blade on a tablesaw **(B)**. Complete the slots by routing each groove with a special T-slot bit in a router **(C)**.

> See "*Tracks and Guide Bars*" on p. 28.

The shank of the bit is guided by the sides of the groove, so you don't need to use an edge guide. Attach the fixture atop your drill press's regular table, using screws, bolts, or cleats. The T-slots not only hold commercially made hold-downs, but you can also use regular T-bolts and nuts or hand knobs to attach wood positioning blocks. When cut to match the profile of your workpiece, these blocks are great for drilling multiple, identical parts.

The vacuum drill-press fixture **(D)**, eliminates the need for clamps by using suction generated by a vacuum valve to firmly hold workpieces during drilling. Make the fixture's table from ¾-in.-thick MDF, sized to be twice as wide as the distance between your drill press's column and chuck center, and as long as you wish. (The one shown here is 14 in. x 18 in.)

To allow use with workpieces of many sizes and shapes, the table has several concentric air passages lined with vacuum "gaskets." Start by drawing a 2-in. square in the center of the table, and then three more rectangles expanding outwards from the center, each one 1 in. to 2 in. larger than the preceding one. These lines designate the locations of thin, sticky-backed neoprene weather-stripping gaskets **(E)**. Before applying the gaskets, rout out a series of shallow air passages, each spaced between the four gasket lines. Fit the router with an edge guide, and a ¼-in. or ⅜-in.-dia. straight bit **(F)**.

Also rout a ⅜-in.-deep x ¾-in.-wide groove along the centerline of the table's length, spanning the innermost and outermost air channels. This groove accepts two 1-in.-long sliding hardwood blocks, cut to fit the groove snugly **(G)**. With a short piece of weather-stripping applied across it, each block acts as a sliding "valve" that directs the vacuum to a smaller or larger part of the fixture, to suit the size and shape of the workpiece. The square area in the middle of the table prevents a loss of vacuum as the bit penetrates the workpiece.

Somewhere along the length of the innermost air passage, groove, bore, and tap a hole for the fitting that connects the vacuum hose to the table.

► See *"Routing Fixtures"* on pp. 222-223.

Mount the table on a pair of support strips that are wide enough to provide clearance under the table for the vacuum hose. (Also add a support block under the table's center to prevent deflection from bit pressure.) Mount the fixture to your drill-press table and attach the vacuum valve in a handy location, for easily engaging and releasing clamping pressure **(H)**.

[TIP] The gasket-enclosed area on a vacuum table must be at least 10 in. sq. to generate enough atmospheric clamping pressure to hold parts securely.

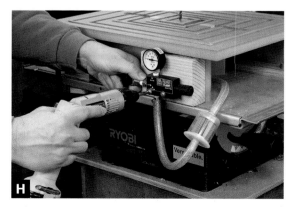

A

B

C

D

Edge-Gluing Fixture

Luthiers often use an edge-gluing fixture like the one shown here to glue thin boards together for a guitar back or cello top. But this fixture is also useful to furniture makers and cabinetmakers, for gluing up door panels, tops for small chests, sides for jewelry boxes, etc.

To make this fixture, first cut a base that's at least 5 in. or 6 in. wider and a few inches longer that the largest panel you plan to create. Use a single piece of ¾-in.-thick melamine, or cover an MDF panel with plastic laminate. The idea is to end up with a flat, sturdy base with a glue-resistant top surface. For bases larger than, say, 3 ft. x 4 ft., you'll need to reinforce the underside with cross braces to keep it flat and true.

Along one long edge of the base, screw a stop strip **(A)** that's at least 1½ in. to 2 in. wide and ¼ in. thicker than the thickest panel you plan to glue up. Use lots of screws, staggering them across the width of the strip to ensure that it is solidly attached and stays flat under clamping pressure. That pressure will be provided by numerous eccentric clamp heads mounted to the opposite side of the base.

Cut the ice-cream-cone-shaped clamp heads out of high-quality plywood **(B)** that's at least as thick as your thickest workpiece. After sanding the clamps' edges smooth, drill a mounting hole through each one that's slightly off-center to the head's large rounded end **(C)**. In this case, the 4-in.-long x 2-in.-wide clamps were drilled ⅛ in. off-center. The offset should be proportionally more for larger eccentric clamps.

Space the clamp heads evenly along the edge of the base, locating them so that their tapered handles overhang the edge **(D)**. Attach the clamps using #10 or #12 wood screws. (For larger clamps, use ¼-in. bolts.) Leave the fasteners loose enough so the clamp heads can be rotated with moderate finger pressure.

Now add two or more cross bars that span the width of the base (E). Attach one end of each bar to the top of the stop strip, and the other to the base between clamping heads, where it won't interfere with clamp adjustment. Fit spacer blocks under each end of the bars to create a gap above the base that's ½ in. wider than the thickest intended workpiece. Use bolts or long wood screws to fasten the bars and spacers to the base. Next, cut a bunch of 4-in. to 6-in.-long wood wedges to use in conjunction with the cross bars to keep boards flat during glue-up.

Before actually gluing up a panel, slide the boards into the fixture, measure the gap left between the panel and the unlocked clamping heads, and cut a spacer strip to this dimension. (For really narrow panels, reposition the clamps on the base.) Mark the order and orientation of the boards, and then apply glue to their edges and set them into the fixture one by one (F). Tighten the cam clamps slightly and then insert wedges under the cross bars wherever they're needed, to keep board edges mated and the assembly flat (G). Tighten the cam clamps (H) and set the whole thing aside to dry. After removing the panel from the fixture, use a plastic scraper or a sponge and warm water to remove the excess glue from the base (I).

[TIP] Dry-clamp a panel assembly before gluing up, to make sure that edges mate snugly and that the clamping process will proceed smoothly.

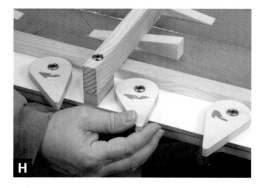

A

B

C

D

E

F

Frame Clamping

A dedicated miter-clamping fixture makes gluing up a batch of picture frames much less of a hassle. Start by cutting a base from melamine, which has a glue-resistant surface. Cut the base 6 in. to 8 in. longer on a side than the frame. Set the dry-assembled, taped-together frame on the center of the base, and nail stop blocks near the corners on two opposite sides **(A)**. Cut two stop strips as long as the other two frame sides, and then cut a scallop at the end of each strip and screw the strip to the fixture base **(B)**. Make four eccentric clamp heads from ¾-in.-thick plywood, and screw them in place so they'll press against the scalloped strip ends.

Set your glued-up and assembled frame into the fixture and turn each eccentric clamp so it keeps each miter joint tight **(C)**. You can use the stop blocks to position the head of a nail gun **(D)**, thus fastening the corners well enough so you can carefully remove the frame, set it aside to dry, and use the fixture to clamp up another one.

You can also build a dedicated fixture for clamping up cabinet door frames and face frames with dowel- or tenon-joined stiles and rails. Cut the fixture's base 6 in. longer on a side than your frame's final dimension. (If you don't have melamine, you can use plywood, applying masking tape to the joint areas.) Lay out the frame's outline with a framing square. Screw stop blocks at the frame's top and bottom edges and at one side edge, using one block at each corner **(E)**. Screw a pair of longer blocks on the frame's opposite side, with each one angled to match the taper of a wood wedge. After gluing and placing the assembled frame members in the fixture, drive in two wedges to firmly clamp the corner joints together **(F)**.

Coopered Panel Fixture

Curved wood doors can make an otherwise plain-looking cabinet a lot more interesting. A coopered panel fixture makes it a lot easier to glue up a series of beveled staves into a perfect curved door, lid, or panel.

Start by calculating the number of staves you'll need for your panel, along with their width and bevel angle. The desired radius of the panel's face determines the radius of the concave form. Cut the forms from 3/4-in.-thick stock, making them at least 6 in. wide and an inch or two longer than the panel's width. Make enough forms to support the panel at least every 8 in. to 12 in. Cut the forms with a jigsaw or an arc-sawing jig **(A)**.

► See *"Arc-Sawing"* on p. 118.

Drill two holes in each form large enough to accommodate the head of a bar clamp **(B)**. Locate the holes on either side of the form's centerline, no more than an inch or so below the curved edge. Cut two 4-in. to 6-in.-wide sides for the fixture from ¾-in.-thick stock, and mark for proper spacing of the forms. Temporarily clamp the sides to the forms, and then screw or nail each one in place **(C)**. To prevent the glued-up panel assembly from sticking, cover each form edge with masking tape **(D)**.

After doing a dry-assembly to ensure that everything will clamp up smoothly, place the glued-up staves in the form and set a couple of scrap strips against the panel's outer edges. Set all the clamps into place first, and then tighten each opposing pair a little at a time, making sure the stave's edges stay aligned and the panel stays straight on the form **(E)**.

[TIP] It's easier to scrape glue drips from a coopered door assembly when the glue is still rubbery than to wait until it's rock-hard.

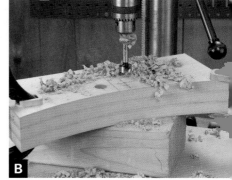

Bending Forms

Wood that's green or that has been heated with steam bends surprisingly easily. But once bent, a part must be held in that position until it has cooled or dried, in order for the curve to be retained. That's where a bending form comes in, to help you bend the stock to shape and hold it there until it sets.

Cut a plywood base for the form that's several inches larger than the final bent part on every side. Draw the desired curve on the base, and then redraw it with a slightly tighter radius. (Bent stock always rebounds a little after it's set.) Depending on the thickness of the workpiece, laminate enough layers of plywood to create a bending form that's at least as thick. Saw out the form's outline curve (in this example, a chair back), and then nail it down on the center of the base **(A)**. Now cut out a few stop blocks from scrap stock and screw them to the baseplate, locating them where the bend in the workpiece begins and ends, as well as somewhere in the middle **(B)**. The stops help hold the work in the form, both as you bend and afterwards. Use a small scrap cut a little thicker than the workpiece as a spacer to help you position the blocks the right distance from the form.

Once you've trimmed your green wood work-piece or steamed it until it's pliable, set one end between the form and one of the end blocks and then pull it gently around the form **(C)**. Continue working it around the form, pressing it down over the blocks to hold it in place. Once the work is entirely in the form, add clamps as necessary **(D)** to press unruly sections flat against the form.

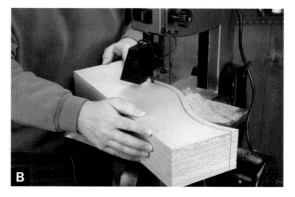

Lamination Forms

By gluing together multiple layers of veneer or other thin wood pieces, you can create curved legs for tables and stools, rockers for rocking chairs, handles for baskets and trays, and much more. One sure way of turning a stack of thin glued laminates into a precise shape is to use a two-part lamination form.

Start by making a cardboard pattern of the outside (or inside) profile of your bent part, drawing curves that account for spring-back in the finished piece. Glue up enough layers of ¾-in.-thick MDF or particleboard to create a block that's slightly thicker and longer than the part, and wide enough to leave at least 4 in. to 6 in. on either side of the curve after it has been cut out. Use your pattern to draw the curve on the block, and then use a compass that is set to the part's final thickness to mark a parallel line **(A)**. On the bandsaw, carefully saw out the section between the two marked lines to create the two halves of your lamination form **(B)**. Clean up the sawn surfaces using a drum or disc sander. To keep the form from sticking together, apply a coat of varnish or polyurethane **(C)**.

Place your unglued stack of wood strips or veneers between the form halves and use as many pipe clamps as necessary to press the forms together. Examine the bent part and, if there are gaps, add tape to shim the form, or sand away high spots as necessary. Remove the workpieces and coat them with glue, using a roller or brush to spread the glue evenly **(D)**.

(Text continued on p. 232.)

Carefully align the edges of the stack and set it between the forms **(E)**. Tighten the clamps below the form slightly, and then add clamps above the form and tighten all of them a little at a time **(F)** until the forms are tightly against the part. Make sure the form halves are aligned and level with each other before leaving the part to dry overnight.

Another way to create parts using the bent lamination method is to clamp the glued-up laminates to a single-sided form. This technique usually requires a lot more clamps than when you're using a two-sided form. However, the bending fixture shown here uses inexpensive ¼-in. bolts to generate the necessary clamping pressure.

Make the form by laying out your desired curve (taking spring-back into account) and then drawing a line that's offset ¾ in. to the concave edge of the curve. This is the pattern to use for cutting out several layers of plywood or particleboard for the form itself. Cut enough layers to glue up a form that's about equal to the width of your part. Using your layout line as a guide, carefully sand the profile of your form on a stationary disc or belt sander until its edge is even and precisely curved **(G)**. To this edge you will attach a series of cross strips on which the laminations will be layed.

Cut the strips from ¾-in.-thick stock that's a couple of inches wider than your form is thick. For the sharper parts of your curve, make the strips ¾ in. wide; for more gentle curved areas, make them up to 1¼ in. wide or wider. Cut enough cross strips to clamp the workpiece every few inches, spacing them closer on sharper curves as necessary.

Now make a clamping strip to match each of the cross strips. On the drill press, bore a ⁵⁄₁₆-in.-dia. hole near each end of each cross strip and its matching clamping strip and then hammer a ¼-in. T-nut into the holes in all the cross strips **(H)**.

Using a small square as a guide, position the cross strips and glue and nail them to the form, with the T-nuts facing down **(I)**. Now put a ¼-in. bolt and washer through the holes in each of the clamping strips.

Do a dry-assembly, clamping your veneers to the form to see that everything fits together correctly. After removing all the clamping strips, apply glue to the veneers and tape the ends of the lamination together to help keep them aligned.

Sandwich the bundle between two strips of ⅛-in.-thick plywood or hardboard to help distribute the clamping pressure of the strips. Set the end of the bundle under a couple of clamping strips at the more tightly curved end of your form (if there is one) **(J)**, and tighten these strips down. As you pull the lamination bundle down over the form, add clamping strips to every other cross strip **(K)**, tightening them only partially. When the bundle is fully bent, add the remaining clamping strips and tighten all of them, using a socket driver in a cordless drill **(L)**. The drill allows you to tighten or loosen the clamping strip bolts much more quickly than you could hand-tighten bar clamps. The drill's clutch also lets you tighten all the bolts to uniform tightness, better ensuring an evenly and cleanly clamped final workpiece.

[TIP] For laminated parts that must conform to an exact curve, it's crucial to glue up a test part first to assess the amount of spring-back after the assembly has dried.

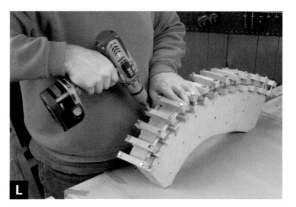

Safety Devices, page 236

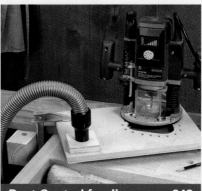

Dust Control for Jigs, page 248

PART FOUR

Safety and Dust Control

WOODWORKING IS SUCH a satisfying and absorbing pastime that it's easy to overlook safety concerns. However, if you work with stationary machines and portable power tools (and who doesn't these days?), safety involves protecting yourself from contact with rapidly spinning blades, cutters, and abrasives. But even if you regularly use power tool guards, you may still be in harm's way because many stock safety devices don't work well with shopmade jigs. To stay safe, you must add protection directly to your jigs and custom setups by adding shopmade guards and hold-downs, and by using special push sticks and blocks when necessary. Another woodshop threat that needs special attention with regard to jigs is fine dust, which can cause lung and sinus damage. With a little innovation, you can fit many jigs with hoods and dust collection ports that collect dust as it's produced, capturing it before it becomes a respiratory hazard.

Safety Devices

Hold-Downs

- ➤ Featherboards (p. 238)
- ➤ Spring Hold-Downs (p. 240)
- ➤ Wheeled Hold-Down (p. 241)

Blade and Cutter Guards

- ➤ Integrated Jig Guards (p. 243)
- ➤ Cutter-Mounted Guards (p. 245)
- ➤ Protective Enclosures (p. 246)
- ➤ Exit Guards (p. 247)

I T'S EASY TO BECOME COMPLACENT about guards when using jigs with your portable power tools and machinery, especially when operations are going smoothly. But when you realize how much protection blade or bit guards, push sticks, hold-downs, and other safety devices afford, and how easy they are to use, there's just no excuse for not taking advantage of them.

Push sticks and blocks are important tools for keeping vulnerable fingers from getting too close (less than 8 to 10 in.) to blades and cutters. The heel of a push stick is used to grab the back edge of the stock to propel it forward (see the bottom photo at left). By making push sticks yourself, from plywood or solid wood, you can customize them to suit your grip and the size of the work. Push blocks are terrific for holding stock flat on a machine's table and are useful on a variety of machines, including the shaper, jointer, and tablesaw. To improve your control of work-pieces, it's a great idea to add a strip of non-skid stair tread tape, such as 3M Safety Walk, to the bottom of your push sticks and push blocks, as shown in the top left photo on the facing page. Keep these devices handy by hanging or storing them on or near your jointer, tablesaw, shaper, and other machines.

Another category of safety devices includes hold-downs such as featherboards,

For maximum safety when working on the tablesaw, shaper, or other aggressive cutting machines, use push sticks and keep hands at least 12 in. away from the cutter.

Non-skid stair tread tape applied to the bottom of a push stick provides greater control over workpieces.

Hold-downs, such as this wheeled shopmade unit that employs skateboard wheels, help keep stock flat on a tool table or against a rip fence while protecting your hands.

spring hold-downs, and wheeled devices. Because they are often fastened very close to the cutter or blade, hold-downs often take the place of guards—preventing dangerous contact with hands. Hold-downs help keep stock flat on a tool table or against a rip fence and can reduce dangerous kickbacks.

Unfortunately, using a jig on a machine tool often requires removing the tool's guard. In these cases, the most practical deterrent to accidents is to use clear plastic guards, enclosures, and/or exit guards. Plastic guards are easy to make using regular woodworking tools. Guards can be screwed to stationary fences and sliding carriages, or even mounted directly to router bits and shaper cutters, as shown in the bottom photo at right. Simple, practical guards can be fitted to nearly all shopmade jigs and setups with very little effort. You'll find that you can also retrofit many of your old jigs and setups to incorporate guards that enhance safety without getting in the way.

The plastic disk mounted to the end of this router bit serves as a guard to protect fingers.

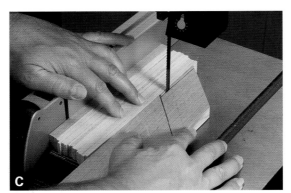

Featherboards

A featherboard is one of the most popular and versatile of hold-down devices. Its flexible fingers apply pressure at an angle against the stock to keep it against a fence or table and to prevent kickback. You can buy ready-made plastic featherboards designed to use with a variety of power machines **(A)**, or you can design your own wooden featherboards to suit your particular needs.

Start with straight-grained, solid wood. Vertical-grain Douglas fir is a good choice because of its consistency and appropriate "springiness." Cut blanks from ¾-in.-thick pieces that are between 2½ in. and 3½ in. wide and at least 9 in. long. Saw one end of each featherboard at a 45-degree angle and then draw a parallel line about 3 in. in from that end to mark the baseline of the fingers.

Thin wood spacer strips are used to guide the finger cuts. They should be evenly spaced and cut thin—usually between ³⁄₃₂ in. and ⁵⁄₃₂ in. thick, depending on the amount of pressure you're after: Thicker fingers apply more force, while thinner fingers are more flexible and apply lighter pressure. Rip each spacer strip to the thickness of the desired finger plus the width of a bandsaw blade kerf **(B)**. Those shown here are ⁵⁄₃₂ in. thick and the same length as the featherboard. Rip as many strips as fingers to be cut.

Stack your spacer strips on edge against the bandsaw rip fence, with the fence set to cut the first finger nearest it. Cut the first finger, sliding the featherboard against your stack of strips **(C)**. Then remove a strip, take another cut, and repeat the process, taking care to stop at the pencil line **(D)**.

You can clamp a featherboard directly to a machine's fence (for ripping, shaping, etc.), or mount it via a pair of slots cut into the featherboard body. The slots can be routed or cut on the tablesaw, using a dado blade to create two parallel, stopped cuts **(E)**. The slotted holes make it easy to properly locate the featherboard and to set the amount of pressure it applies to the stock.

To mount a wood featherboard flat on a machine table, you can make the featherboard long enough to clamp at the table's edge, or you can attach it with magnets. Alternatively, you can screw it to a scrap of wood that clamps to the table, or use T-slot fasteners to attach it directly to T-slot-style miter slots. One flexible approach involves clamping a wood strip into your machine's miter gauge slot and then attaching a featherboard to the strip **(F)**.

For certain cuts, featherboards are often used in pairs. A featherboard or spring hold-down mounted on the fence keeps stock flat on the machine table, while another mounted on the machine table keeps the stock firmly riding against the fence **(G)**. This works well when you're cutting straight-edged strips and parts of all sizes. Usually, it's best to press the featherboard against the stock near the point of the cut. When the workpiece has a curved or irregular edge, use a pair of featherboards, with one pressed freehand **(H)** against the irregular side of the work.

[TIP] To mount a featherboard to an iron or steel table/fence quickly and easily, try inlaying two or more rare-earth magnets into its body, using epoxy or cyanoacrylate glue.

⚠️ **WARNING Never position a featherboard to press stock directly against the side of a blade or cutter because it can pinch the stock, causing kickback.**

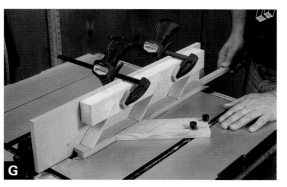

Spring Hold-Downs

Spring hold-downs are simple devices that use a strip of wood or steel to apply pressure to a workpiece to keep it snug against a fence or table during cutting. Commercially made spring hold-downs typically incorporate one or two flexible, adjustable steel arms **(A)**. But you can make a simple spring hold-down with wood slats. Start by cutting several ⅛-in.-thick strips from a resilient wood like oak, ash, or vertical-grain Douglas fir. I cut the fir slats for my general-purpose hold-down ⅛ in. thick and 4½ in. long, but your slat's dimensions depend on the wood you use (some woods are stiffer or springier than others) and how much pressure you wish to apply. (Lighter cuts and thinner stock require less pressure than heavy cuts in thick workpieces.)

Make the slat holder by cutting a section of 2x2 stock to about 18 in. long. Next, tilt your tablesaw for a 45-degree cut and use a standard blade that creates a ⅛-in. kerf to saw equally spaced slots for the slats **(B)**. Before gluing and tacking the slats into their slots, bevel the bottom edge of each slat slightly, using a block plane **(C)**. Orient the beveled edges so they're parallel with the holder, and drive a couple of nails into the holder to secure each slat.

Attach the slat hold-down to your machine table or fence with clamps **(D)**, or if your fence has a T-track, mount them with ¼-in. T-bolts. It's best to locate the hold-down so that the center slat is just ahead of the cutter/blade, with one slat before and one after the cut. To set the hold-down for a cut, bring it in light contact with the workpiece, so that the slats are deflected slightly.

Wheeled Hold-Down

Commercial wheeled devices feature a spring-loaded arm that presses the wheel down to keep stock flat on the table. These devices are usually sold and used in pairs, and mounted so that one is located just ahead of the bit or blade, and one after it. The wheels angle slightly, relative to the line of cut, pulling the stock snug against the fence as it is fed. A ratchet mechanism inside most brands of wheels allows only forward feeding, preventing the stock from kicking back should it bind on the blade or bit. Some units clamp to the top of the fence **(A)**, while others mount to T-track slots found on rip fences like the one made by Excalibur.

You can also build a versatile wheeled hold-down from a few scraps of plywood, a couple of springs or rubber bands, bolts and screws, and urethane plastic skateboard wheels. Although it won't provide kickback protection, this pivoting-arm wheeled hold-down is useful with many different tools and jig setups. By attaching the arm to a base that clamps to your tablesaw's rip fence rail, you can position the arm to hold stock against the tablesaw when you're ripping **(B)**. You can also mount it to a shaper to hold stock against the fence when cutting moldings. The pivoting arm can also be mounted to a magnetic base and used to keep stock firmly against your bandsaw's fence during resawing.

To build the basic pivot arm hold-down, cut out the pivot arm and base from ½-in.-thick Baltic birch ply-

(Text continued on p. 242.)

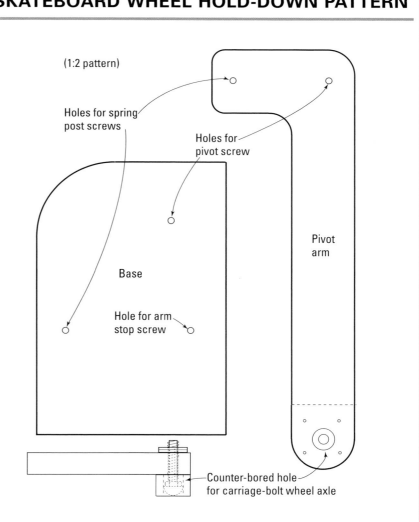

SKATEBOARD WHEEL HOLD-DOWN PATTERN

(1:2 pattern)

Holes for spring post screws

Holes for pivot screw

Pivot arm

Base

Hole for arm stop screw

Counter-bored hole for carriage-bolt wheel axle

SKATEBOARD WHEEL HOLD-DOWN

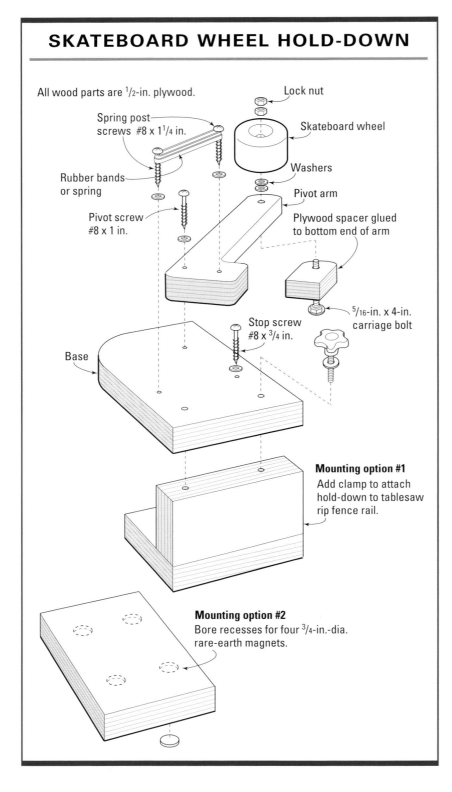

All wood parts are 1/2-in. plywood.

Lock nut

Spring post screws #8 x 1¼ in.

Skateboard wheel

Rubber bands or spring

Washers

Pivot screw #8 x 1 in.

Pivot arm

Plywood spacer glued to bottom end of arm

Stop screw #8 x ¾ in.

⁵/₁₆-in. x 4-in. carriage bolt

Base

Mounting option #1
Add clamp to attach hold-down to tablesaw rip fence rail.

Mounting option #2
Bore recesses for four ³/₄-in.-dia. rare-earth magnets.

wood as shown in the drawing on p. 241. Drill holes through both the arm and base and attach the arm with a #8 x 1-in. round-head wood screw and washer as a pivot for the arm. As shown in the drawing at left, another #8 wood screw driven into the base acts as a stop for the arm. Two more screws—one on the base and one on the bent part of the pivot arm—provide attachment points for the bands or springs that tension the arm.

Skateboard wheels (available at sport and skate shops) come in sets of two or four wheels. Each wheel requires two sealed ball bearings, which are pressed into wheels with an arbor press (most skate shops do this for free). Although the standard skateboard bearing I.D. is 8mm, a regular ⁵/₁₆-in. carriage bolt works well as an axle to mount the wheel (or wheels) on the end of the pivot arm **(C)**.

When you're cutting thin strips or when lighter hold-down pressure is adequate, several wide rubber bands looped over the posts can exert up to 4 or 5 pounds of pressure on the arm **(D)**. For heavier work, switch to a pair of two ⁵/₁₆-in.-dia. by 1¼-in.-long extension springs.

Integrated Jig Guards

For greatest woodworking safety, jigs and fixtures for power tools should be fitted with
guards or protective devices made from clear
polycarbonate plastic. Sold in ⅛-in.-thick sheets at
home centers under the name Lexan, this plastic
is shatterproof, easy to work with, and transparent enough for good visibility while working. It is
easily cut on a tablesaw with a fine-tooth blade
or any blade made for sawing laminate materials
such as plywood **(A)**. Mounting holes for guards
are easy to drill with regular woodworking bits.
(Drill holes larger than 1 in. in diameter using
multi-spur bits or hole saws, as Forstner bits tend
to grab.) Mounting slots should be routed **(B)**
with spiral-flute bits, as regular straight bits tend
to chew up the plastic.

▶ See *"Clear Plastics"* on p. 19.

Lexan is relatively easy to bend when heated.
Use a heat gun (the kind for removing paint) to
soften the thin sheet stock. To make a clean
90-degree or sharp-angled bend, cover the portions of the plastic adjacent to the bend by
clamping on small wood scraps **(C)**. Move the
heat gun over the uncovered portion slowly, until
the plastic becomes soft enough to bend easily.
Hold the piece in the bent position until it cools.
As we'll see, you can even bend Lexan into
curved shapes for more elaborate guards.

Most jigs can be fitted with some kind of guard
that prevents fingers from contacting a blade or
cutter, although some types of jigs are easier to
fit than others. To add a guard to a sliding miter
jig, screw a strip of Lexan to the jig, mounted
directly over the line of cut **(D)**.

(Text continues on p. 244.)

You can make a guard for a sliding tenon jig guard by attaching a Lexan strip to the side of the jig over the blade **(E)**.

Most commercially made router tables come with a fence-mounted plastic cutter guard **(F)**. To make your own router table or shaper guard, take a rectangular strip of Lexan and make one end semi-circular using a jigsaw fitted with a fine-tooth blade. Rout a pair of slotted holes in the square end (to allow up-and-down adjustment), and mount the guard to a T-track in the fence, for horizontal adjustment.

To add a guard to a tablesaw box joint jig, a frame must be added to the jig to support the thin Lexan sheet above the blade **(G)**. The solid-wood frame is screwed to the back side of the jig, with the Lexan piece screwed to the top of the frame. (In this photo, the edge of the Lexan is being marked with a felt-tip pen, for better visibility.) A space between the back edge of the Lexan and the support face of the jig allows the workpiece to be inserted for cutting. In addition to preventing accidental contact with the blade, the Lexan also deflects chips thrown upward.

One way to make a sliding crosscut carriage safer is to fit it with a tunnel-like guard directly above the line of cut **(H)**. Make the three-sided guard from two thin strips of wood capped by a long, narrow Lexan strip screwed to the top. Slots and guide strips on the jig's fences allow the guard to slide up and down easily. In use, the workpiece is placed in the carriage, and then the guard is slipped on top of it before the cut is made.

> **⚠ WARNING** Never use acrylic plastic (like Plexiglas) for woodworking guards because it can shatter upon impact. Instead, use polycarbonate plastic, as discussed on p. 19.

Cutter-Mounted Guards

Lexan can also be put to good use making guards that mount directly to router and shaper bits and cutters. A round, cutter-mounted guard is not only easy to make and use, but it's practically the only way to protect yourself from serious injury when you're freehand shaping and template routing **(A)**. For heavy-duty shaper work, clear guards with built-in bearings sized to fit standard spindles are de rigueur **(B)**. It's foolhardy to work without this kind of guard any time the shaper's usual fence and guard must be removed for special operations.

Overhead guards for router bits aren't readily available but are easy to make from ⅛-in.- or ¼-in.-thick clear Lexan. All it takes to cut one out is a hole saw mounted in the drill press **(C)**. The hole saw should be large enough to create a guard at least ½ in. larger in diameter than the bit it'll be used with. Use a hole saw with an interchangeable pilot drill, one whose diameter matches the router bit's bearing screw or one that fits atop the threaded shaft of an arbor-style router cutter **(D)**. Since the guard doesn't spin independent of the bit like a shaper guard, you definitely need to sand the outer rim of the guard smooth and round. That way, if your hand accidentally makes contact during routing, the worst you'll get is a burn. To help you see and avoid the transparent guard, use a felt-tip pen to color the edges bright red.

> ⚠ **WARNING** Even with a guard mounted, keep your hands away from large shaper cutters or router bits. If the cutter suddenly grabs the stock, your fingers could be pulled into the bit.

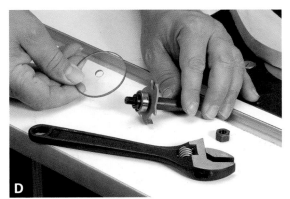

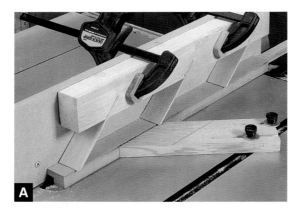

A

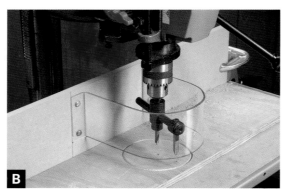

B

C

D

Protective Enclosures

Hold-downs and featherboards used with fences and sliding jigs can provide plenty of user protection, since they completely cover the blade area **(A)**. You can also use Lexan (polycarbonate) clear plastic to create custom enclosures that provide good protection from blades, bits, and abrasives.

A good example of this kind of protective enclosure is the drill-press guard shown in photo **(B)**. The guard protects your knuckles from getting banged or ripped up by a large adjustable protractor bit. These big bits are designed to spin relatively slowly and cut large holes in sheet materials. The guard is made from a single piece of ⅛-in.-thick clear Lexan that's been heated and bent into a semi-circular shape and mounted to an auxiliary face attached to the drill-press fence.

To get a clean, consistent curve, start by cutting a half-round form from a piece of 4x6 construction lumber. Alternatively, you can use a coffee can, pipe, or other rigid object for a form. After cutting a strip of Lexan long enough for the desired guard (this one is 18 in. long), drill two mounting holes at each end. Clamp one end of the strip to the side of your form and heat the plastic with a heat gun until it's flexible enough to bend by hand **(C)**. Hold the bent plastic down on the form until it cools, retaining its curved shape. Next, bend the ends of the strip into sharp 90-degree flanges using the procedure described on p. 243. Finally, use screws to attach the guard to your drill-press fence (or wood auxiliary fence face, as shown here) **(D)**.

➤ See *"Integrated Jig Guards"* on p. 243.

> ⚠ **WARNING** Whenever you use a heat gun to bend plastic, wear thick leather work gloves to protect your hand and prevent an accidental burn.

Exit Guards

Another way to protect precious digits is to add exit guards to sliding jigs. With most sliding jigs, the blade or cutter is exposed at the end of the cut, which is often when you're reaching over the top of it. An exit guard is nothing more than a block that completely encloses the blade or cutter at the end of the cut **(A)**.

It's pretty easy to fit an exit guard to most jigs, simply by gluing and screwing a thick wood block to the jig's baseplate or rear rail. You can make one from any thick scrap or construction lumber cutoff. It's best to locate the block on the jig just past the area of the final cut. For example, on a sliding miter jig, a triangular exit block is located behind the intersection of the two fences **(B)**. On a sliding crosscut sled or carriage, the exit guard block is easy enough to screw directly to the jig's rear rail, centering it on the blade **(C)**. Make the block wide enough so that the blade remains enclosed when the jig has been moved forward as far as is necessary to complete the cut. Alternatively, you can use plywood to build an exit guard that's like a hollow box. Made large enough to enclose the blade or cutter, this guard box also serves double duty as a dust collection shroud **(D)**. Before use, clamp a stop to your machine that positively prevents the jig from sliding far enough for the blade or cutter to pass through the exit guard.

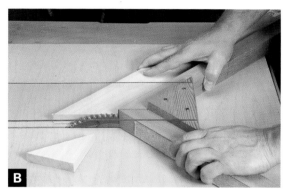

Dust Control for Jigs

Jig Dust Collection

Power Tool Pickups

WHIRLING CUTTERS AREN'T the only danger to woodworkers in the shop; sawdust is also a threat. Piles of sawdust are a fire hazard, and fine, floating dust particles can trigger allergic reactions, and cause lung damage or even cancer. Even if most of your shop's machines are connected to a central dust system, collection can suffer: Stock guards often must be removed from machines like tablesaws and shapers for the jigs to function properly, and portable power tool jigs may thwart the tool's integral dust collection. Fortunately, there are ways to incorporate dust collection into jigs and fixtures used with your power tools.

The simplest way to control dust when using a custom jig setup is to position a vacuum hose near the area of dust ejection by a blade, bit, or abrasive disc, or belt. In some cases, you can get away with temporarily clamping, strapping, or duct-taping the hose itself to the machine tool table or the jig. To improve the capture of dust and chips over a broader area, it's better to fit a dust shroud or hood to the end of the hose. A dust hood—typically shaped like a nozzle, funnel, or box—is designed to deflect chips towards the vacuum hose opening, as shown in the top photo on the facing page.

An even better way to deal with power tool dust is to build dust collection directly into the jigs you make. In addition to chip pickup, a hood also provides a convenient place to mount a hold-down, clear guard, or other safety device, and can offer other advantages as well. (See sidebar below.)

You can also incorporate dust collection into table-style jigs and fixtures by building hollow box-like tables that connect to a shop vac or dust collector. This kind of downdraft table removes dust created by many different operations, including mortising, routing or shaping, and sanding, as shown in the top photo on p. 250.

Many modern portable power tools come with built-in hose ports for attaching dust collection hoses. However, when these tools are used with jigs, built-in dust collection is

Power sanders create lots of fine dust that's easiest to collect near the source. In this case a collection hood located close to the sanding disc does the trick.

▶ INTEGRATED DUST HOODS

A dust collection hood built into a jig or fixture does a good job of keeping your shop—and lungs—cleaner and healthier. But it offers other advantages as well. A dust hood increases safety by covering or surrounding a cutter or blade and deflecting chips that might otherwise fly up in your face. Sucking chips away from the cutter can even improve the quality of your cuts and extend the life of your blades and cutters: Chips that aren't vacuumed away after the initial cut may be recut many times as they tumble about, increasing the wear on cutting edges. Finally, collecting dust and chips before they can fall into a tool's motor or working parts can prolong

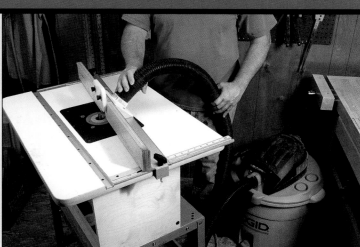

the life of the machine. This is an especially important consideration with routers used in router tables, since most units aren't designed to run in an inverted position and can actually be damaged by chips entering the lower motor housing.

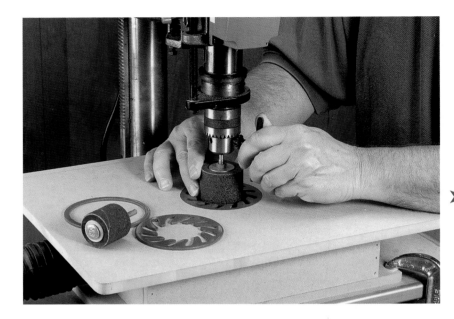

The hollow-box construction of this drum sanding table allows connection of a vacuum hose to capture the fine dust as it's created.

Adding dust collection to a template not only makes routing cleaner but keeps chips from clogging the template and interfering with routing.

often defeated or hindered. For example, outfitting a router for dust collection is notoriously difficult, even when the tool is used for fairly simple jobs like edge routing or mortising. But you can devise a custom dust pickup that mounts to the router to capture the lion's share of chips when edge routing.

► See *"Router Dust Pickup"* on p. 258.

You can also integrate dust collection into most router jigs. A good example is a basic routing template used for creating inlays, recesses for hardware, or even for cutting joinery. A special hollow template with air passages removes dust around the template opening and whisks it away through a shop vacuum hose (see the photo below).

Sliding Dust Pickup

Because most crosscut sleds, miter-cutting jigs, tenon-cutting jigs, and other tablesaw jigs pass directly over the blade, the saw's stock blade guard must usually be removed. In these cases, it's best to incorporate a custom-made blade guard into the jig itself. Better yet, you can build dust collection into your custom guard by incorporating an overhead dust pickup.

Adding a dust pickup to a jig such as the miter sled shown **(A)** is easy. First strengthen the thin plastic guard to support a vacuum hose port. Add a ½-in. by ½-in. wood strip to each edge of the plastic **(B)**. You can either cut a groove along the length of each strip to capture the edge of the plastic, or cut a rabbet on the strip and screw it to the guard **(C)**. The easiest way to attach a vacuum hose is to add a ready-made hose port (available from woodworking supply catalogs), which attaches over a 2¼-in.-dia. hole bored through the guard. It is fastened with four small machine screws and washers **(D)**. Locate the port near the middle of the guard to capture debris thrown upward by the blade.

To add dust collection to a tablesaw tenon jig, bore a hole through the jig's guard **(E)** to provide a snug mount for a small vacuum hose. Use a circle cutter, hole saw, or Forstner bit chucked in the drill press to bore the hole near the rearmost section of the guard. To keep the hose from interfering with the sliding jig during use, secure the hose to the jig with a plastic pipe strap, locking tie, or twist tie.

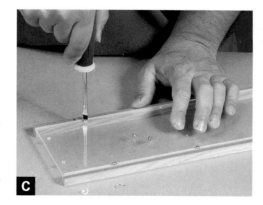

[TIP] Dust collection works best when chips are evacuated in the same direction they're thrown by the cutter. Observe the travel of debris to best locate the dust port.

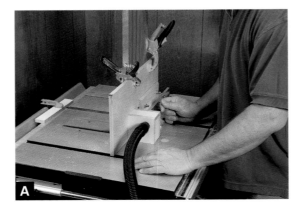

Exit Guard Dust Box

An exit guard protects you by keeping blades and cutters sheathed at the end of a cut. However, you can make it do double duty by also employing it as a small dust port box, as seen on this finger joint jig **(A)**. The dust port box is made from thin plywood and attaches to the rear wall of the jig. The dust port hole leads out of the side of the dust box, so the hose won't get in the way during use.

Begin by making the four sides of the box, which has an open bottom and front. Bore a hole in one of the side pieces using a hole saw or Forstner bit **(B)**. Glue the sides of the box to its top and back pieces, reinforcing the butt-joined parts with brads. For easier assembly, apply glue to the plywood edges and then tape the box together **(C)** before driving the brads.

Mount the box to the back side of the jig, centering it in the blade's path. Because the bottom of the box slides directly on the saw table, it's easiest to attach the box with the jig setting in place on the saw. A block of scrap wood temporarily clamped to the edge of the saw keeps the jig from sliding as you attach the dust box with four #6 brass screws **(D)**.

> ⚠ **WARNING** Never use any sliding jig without first clamping a positive stop to the machine to prevent the jig from sliding far enough to expose the blade or cutter.

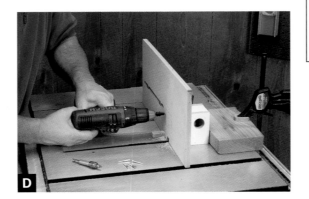

Integrated Dust Hood

Part-shaping jigs used on the router table or shaper often require the removal of the machine's stock fence and dust hood or pickup. Fortunately, it usually doesn't take much effort to incorporate a new dust pickup hood directly into the jig itself. A good example is the biscuit joiner guide discussed on p. 148. In this case, a simple plywood box provides an adequate dust hood and attachment for a vacuum hose. The dust hood not only allows chip collection when you're routing biscuit joinery slots, but it also provides ideal mounting for a clear plastic guard that protects the user from the bit.

As with other hoods discussed in this section, this one for the biscuit jig is glued and nailed together from ½-in.-thick plywood scraps. The width of the hood is sized to fit a "crevice tool," an accessory nozzle that comes as a standard or optional accessory for many shop vacs **(A)**. The height of the three-sided hood is determined by the cutting height of the jig's kerf-cutting bit. In this case, the sides are about 1¼ in. high.

After the box is assembled, a clear Lexan guard is screwed to the underside of the hood's top piece **(B)**. To add a little friction to prevent the crevice tool from slipping out of the hood, a small piece of dense foam weather-stripping is adhered to the inside of the box **(C)**. After positioning the hood on the jig platform (making sure that it clears the bit and the jig's pivoting arms), attach it with a couple of #8 x 1¾-in. screws **(D)**.

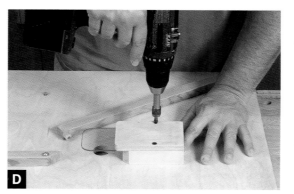

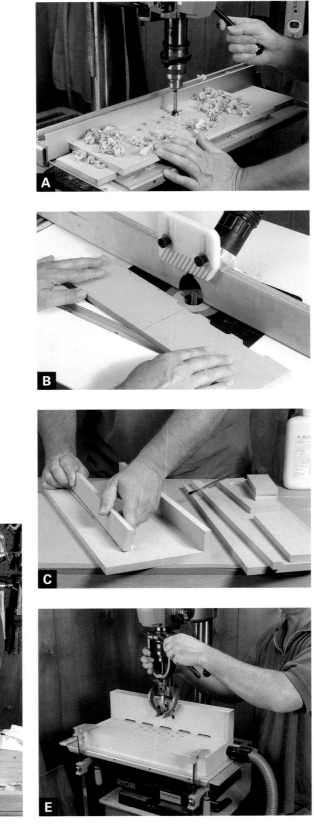

Dust-Collecting Fence

When you're boring holes or mortising frame stock on the drill press, chips can prevent the stock from bearing against the fence, resulting in misplaced holes or mortises. A special dust-collecting fence can help. The fence is basically a hollow, L-shaped box built from ½ in.-thick MDF. Holes in the table and fence allow chips to be sucked away through a vacuum hose in the side of the box. Staggering ⅝-in.-dia. holes in a pattern as shown allows plenty of holes without weakening the pieces too much **(A)**.

To prevent chips from piling up at the juncture of fence and table, rout slots along the lower edge of the fence. Do this on the router table with a ½-in.-dia. straight bit projecting ¼ in. out from the face of the fence **(B)**.

Assemble the fence by gluing and nailing the front and a support block to the baseplate **(C)**. The support block prevents the jig's table from deflecting in use. Glue and nail the fence to the top of the table at its rear edge, and then attach this sub-assembly to the front edge strip and support block, aligning all edges and ends. Glue the back and top strip on, and then the end pieces, having drilled a hole in one of them to accommodate a shop vacuum hose **(D)**. For good suction, ensure that all edges fit snugly together and that all joints get a good coating of glue.

In use, clamp the fence to your drill press or mortiser table **(E)**. When mortising, you'll have to employ a hold-down to keep the workpiece from lifting when you're retracting the chisel.

Drum-Sanding Dust Table

A drum sander chucked in a drill press is a very handy tool indeed, especially if you have lots of curved parts to smooth. It's an even better setup, though, if you build a special drum-sanding table to capture fine dust and whisk it away before it fills your lungs and coats your benches and countertops.

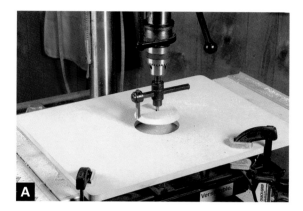

This dust table is a hollow, perforated box that clamps to your drill-press table and connects via a hose to your shop vacuum or central dust system. You can build it from ½-in.-thick MDF or plywood pieces, glued and nailed or screwed together. Cut the parts out on the tablesaw and then drill a hole for your sanding drum using a hole saw or circle cutter **(A)**. If you plan to use only a single drum, make the hole ¼ in. larger in diameter than the drum. But here's a neat trick to make the table more versatile: First, bore a 3⅜-in.-dia. hole through the top. Then use a rabbet bit and a router to create a ⅛ in.-deep, ¼-in.-wide rabbet all around the opening **(B)**. This adapts the top to accept throat plate inserts from a Delta™ oscillating spindle sander. By swapping throat plates, you can use the table with several different sizes of sanding drums **(C)**.

After cutting all the parts to size, bore a hole in one of the side frame pieces to fit your vacuum hose. Assemble the dust table by gluing and butt-joining the sides together, and then attaching the top and base **(D)** and using nails or screws to secure parts together. Notice that the table's base is longer than the width of its frame, creating a 1½ in.-wide clamping lip at each end of the unit **(E)**.

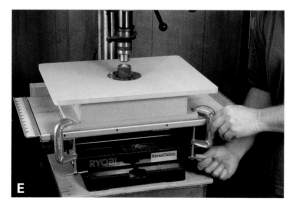

DUST-EXTRACTING ROUTER TEMPLATE

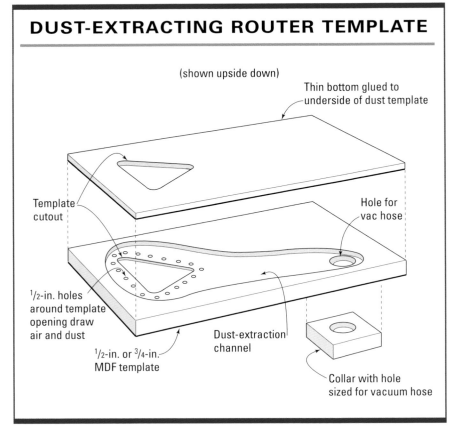

(shown upside down)

Thin bottom glued to underside of dust template

Template cutout

Hole for vac hose

1/2-in. holes around template opening draw air and dust

1/2-in. or 3/4-in. MDF template

Dust-extraction channel

Collar with hole sized for vacuum hose

Dust-Extracting Router Template

Using a bushing-guided router with a template is a powerful method for creating recesses for hardware, decorative inlays, or even for mortising stock. However, the prodigious amount of dust and chips produced by routing is both annoying and health-threatening. The chips are also prone to clogging the template opening, which can prevent the guide bushing from properly contacting the template. A very elegant solution to this problem is to create a dust-extracting router template that uses the suction power of an ordinary shop vacuum to clear away chips and dust.

A dust-extracting template simply consists of two parts: A ½-in. or ¾-in.-thick MDF top and a ⅛-in. or ¼ in.-in.-thick hardboard bottom. The top piece is cut out to serve as a template against which the bushing rides. As shown in the drawing, it also includes dust passages and a port for the vacuum hose. The bottom piece simply covers the dust passages to improve suction.

To create the top piece, cut a rectangular piece of MDF that's wide enough to include at least a few inches all the way around the template cutout to support the base of the router. The top piece should be long enough to allow room for a vacuum hose connection in one corner without impeding router movement in use. Cut out the template hole using a jigsaw or coping saw **(A)**, sizing the opening to include the router bushing offset, as shown on p. 170. Drill a dust port hole that matches the diameter of the vacuum hose you intend to use, and then make a mounting collar for the hose from a square piece of ¾-in.-thick scrap drilled to accept the vacuum hose connector.

The dust extraction channel is routed out from the underside using a ½-in.-dia. straight bit **(B)**.

A

B

Set the router's depth of cut to leave at least ¼ in. of thickness around the template opening to support the guide bushing. Rout at least ¾ in. back from the template's opening all the way around, and then rout a wide passage leading to the dust port hole.

Flip the template right-side up and glue and nail the vacuum collar in place over the port hole **(C)**. Next, bore a row of ½-in.-dia. holes all the way around the template opening **(D)**. These air-inlet holes keep the shop vac from straining and help collect dust from the top of the template, keeping it cleaner.

Make the bottom from a piece of thin hardboard the same width and length as the top. Align the two pieces, trace the template opening onto the bottom piece, and then cut it out with a jigsaw. Nail the bottom to the underside of the top after applying a liberal coat of glue to seal the connection **(E)**.

In preparation for routing, the dust-extracting template is firmly clamped atop the workpiece **(F)**, and the vacuum hose is pushed into the port. Fit your router with the guide bushing and straight bit appropriate to the template's offset and you're ready to rout **(G)**.

▶ See *"Calculating Template Offset"* on p. 170.

[**TIP**] To prevent a dust-extracting template from slipping out of position during routing, apply a few strips of fine-grit PSA-backed sandpaper to the underside of the jig before clamping it down.

C

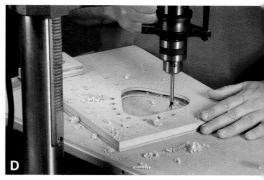
D

E

F

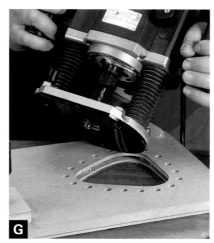
G

Router Dust Pickup

A router is one of the workshop's biggest producers of dust and chips. However, you can squelch some of the dust clouds by using a shopmade dust pickup designed to collect dust and chips thrown off when you're routing edges **(A)**.

The pickup is constructed from a few small pieces of plywood and hardboard. First, make a pattern for the pickup sub-base by tracing the profile of your router's sub-base onto a piece of stiff paper or thin cardboard **(B)**. On the paper, sketch in the V-shaped outline of the dust pickup, sized and proportioned to suit your router. Cut out the pattern and adhere it to a piece of ⅛-in.- or ¼-in.-thick hardboard or Lexan, using spray adhesive or adhesive transfer tape. Cut out the pickup sub-base using either a jigsaw or a bandsaw **(C)**.

To mark out the pickup's mounting holes, set the router's stock sub-base bottom-side up on the underside of the piece you just cut out. Using what's left of your pattern as a guide, rotate the sub-base so that the pickup's V-shaped hood ends up in the desired orientation relative to the router's handles. Use the sub-base's mounting holes as a guide for locating the mounting holes in the pickup sub-base. Then drill the holes using a self-centering drill bit **(D)**.

After cutting out the pickup's sides and bottom, drill a vacuum hose hole in one of the sides, and then glue and nail the two sides to each other to form the V. Attach this assembly to the pickup sub-base and glue on the triangular bottom cover **(E)**. If necessary, drill a hole in the bottom cover to access one of the sub-base mounting screws when you're attaching or removing the dust pickup.

Hose Tether

Many newer-model circular saws, sanders, routers, and other portable power tools have integral dust collection ports for connection to a shop vac hose. Unfortunately, an attached hose can make a power tool feel terribly unwieldy. A tether attached to your bench or work counter can help keep a hose out of your way.

You can tack together a hose tether in minutes from a couple of pieces of 1x2 stock. Cut a post 36 in. to 42 in. long, and nail a 12-in.-long arm to one end. Reinforce the joint by gluing and nailing on a couple of gussets cut from thin plywood scrap **(A)**. Install a screw eye near the end of the arm to hold a short bungee cord that will support your power tool's hose **(B)**. If the tether will be held in a bench vise, nail on a cleat a few inches from the end to make the post sit more securely.

If you dedicate a vacuum hose to a single power tool, attach the tool's electrical cord to the hose with duct tape or wire or plastic ties **(C)**. Attaching the two together makes either less likely to snag or hang up as you work. Loop the bungee cord around the hose and cord and attach it securely to the screw eye so it won't spring back suddenly and cause injury. Be sure to leave enough slack in the hose to allow it to reach to the end of a long board you might be working on **(D)**.

[TIP] If your shop has low, open rafters, you can simply attach a bungee cord to a rafter above your bench to hang a portable power tool's cord and hose.

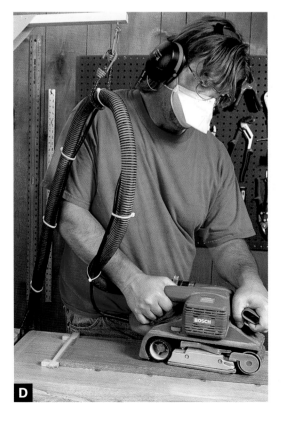

Index

Other Books in the Series:

HARDCOVER

The Complete Illustrated Guide to Joinery
Gary Rogowski
ISBN 1-56158-401-0
Product #070535
$39.95 U.S.
$54.95 Canada

The Complete Illustrated Guide to Furniture and Cabinet Construction
Andy Rae
ISBN 1-56158-402-9
Product #070534
$39.95 U.S.
$54.95 Canada

The Complete Illustrated Guide to Shaping Wood
Lonnie Bird
ISBN 1-56158-400-2
Product #070533
$39.95 U.S.
$54.95 Canada

Taunton's Complete Illustrated Guide to Finishing
Jeff Jewitt
ISBN 1-56158-592-0
Product #070712
$39.95 U.S.
$54.95 Canada

Taunton's Complete Illustrated Guide to Sharpening
Thomas Lie-Nielsen
ISBN 1-56158-657-9
Product #070737
$39.95 U.S.
$54.95 Canada

Taunton's Complete Illustrated Guide to Using Woodworking Tools
Lonnie Bird
ISBN 1-56158-597-1
Product #070729
$39.95 U.S.
$54.95 Canada

Taunton's Complete Illustrated Guide to Turning
Richard Raffan
ISBN 1-56158-672-2
Product #070757
$39.95 U.S.
$54.95 Canada

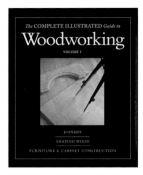

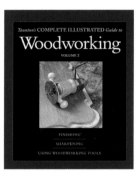

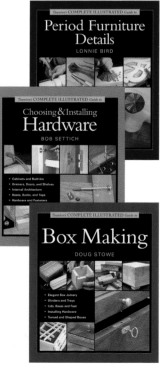

THE COMPLETE ILLUSTRATED GUIDES SLIPCASE SET VOLUME 1

The Complete Illustrated Guide to Joinery

The Complete Illustrated Guide to Furniture and Cabinet Construction

The Complete Illustrated Guide to Shaping Wood
ISBN 1-56158-602-1
Product #070665
$120.00 U.S.
$170.00 Canada

TAUNTON'S COMPLETE ILLUSTRATED GUIDES SLIPCASE SET VOLUME 2

Taunton's Complete Illustrated Guide to Using Woodworking Tools

Taunton's Complete Illustrated Guide to Sharpening

Taunton's Complete Illustrated Guide to Finishing
ISBN 1-56158-745-1
Product #070817
$120.00 U.S.
$170.00 Canada

PAPERBACK

Taunton's Complete Illustrated Guide to Period Furniture Details

Lonnie Bird
ISBN 1-56158-590-4
Product #070708
$27.00 U.S.
$38.00 Canada

Taunton's Complete Illustrated Guide to Choosing and Installing Hardware

Bob Settich
ISBN 1-56158-561-0
Product #070647
$29.95 U.S.
$42.00 Canada

Taunton's Complete Illustrated Guide to Box Making
Doug Stowe
ISBN 1-56158-593-9
Product #070721
$24.95 U.S.
$34.95 Canada